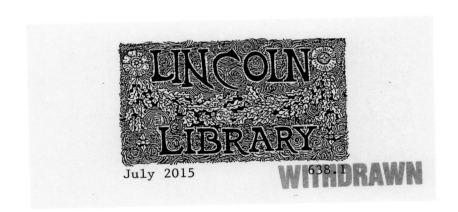

# ~ THE ~
# BEEKEEPER'S
## PROBLEM SOLVER

© Quid Publishing 2015

First published in the United States of America in 2015 by
Quarry Books, a member of
Quarto Publishing Group USA Inc.
100 Cummings Center
Suite 406-L
Beverly, Massachusetts 01915-6101
Telephone: (978) 282-9590
Fax: (978) 283-2742
www.quarrybooks.com

10 9 8 7 6 5 4 3 2 1

ISBN: 978-1-63159-035-1

Conceived, designed and produced by
Quid Publishing
Level 4, Sheridan House
114 Western Road,
Hove, BN3 1DD
England
www.quidpublishing.com

Designed by Clare Barber
Written by James E. Tew

Printed in China

# THE
# BEEKEEPER'S
## PROBLEM SOLVER

*James E. Tew*

**Quarry Books**
100 Cummings Center, Suite 406L
Beverly, MA 01915

quarrybooks.com • craftside.typepad.com

# CONTENTS

# INTRODUCTION

I f practiced properly, beekeeping is an enjoyable and beneficial endeavor—both for the beekeeper and the bees. A healthy hive during warm summer months is a powerhouse of efficiency and productivity that is always a pleasure to watch. The air in the apiary is filled with the sound of thousands of airborne foragers and the floral odor of newly gathered nectar. New beehive equipment, with its own pleasing smells, is purchased and assembled with the anticipation of it being filled with large honey crops and gentle bees that pollinate blossoms all around the community.

This short list is only a small sample of the many positive aspects of beekeeping. However, as with any meaningful craft, problems can arise. Indeed, they should even be expected. Managing the challenges presented by an ailing colony cannot be considered pleasurable, but resolving them successfully is certainly fulfilling. No matter the bee colony's problem, there is always a way that the determined beekeeper can help to improve matters. Bees are resilient and, as a beekeeper, so must you be.

The beekeeping craft can be practiced from a small domestic level up to a large commercial operation. For those who keep bees, the craft is challenging and diverse enough to provide growth and enjoyment for a lifetime. This book discusses 100 of the most commonly encountered problems in beekeeping and provides practical solutions. There are issues that apply to all levels of experience, from the beekeeping basics in chapter one to more advanced practices, such as the queen-grafting techniques covered in chapter five. The essentials of beekeeping—beekeeping equipment, hive management and bee biology—are all covered here, and a chapter is devoted to the pests and diseases known to thwart honey bees. Consideration is also given to the collection and processing of honey and there is also mention of other by-products such as beeswax.

For the keen beekeeper who wants to expand their knowledge even further, or if you don't find your specific concern addressed within these pages, the further resources section will direct you on the right path. But let it be acknowledged that, although humans have been keeping bees for hundreds of years, there are still aspects of bee biology and the life of the colony that remain a mystery to beekeepers and scientists alike.

You can either dip in and out of this book, referring to specific problems as and when they become relevant to you, or you can read the entire collection in one go. Whichever approach you choose, I hope that the knowledge you gain will enhance your experience, and that you enjoy many years of fruitful beekeeping.

## CHAPTER ONE
# BEEKEEPING BASICS

People have been keeping bees for hundreds of years in nearly all parts of the world. The crops of honey and wax to be harvested from managed hives have always been the primary motivation for the practice, but today there are many other reasons why people choose to keep bees. Though beekeeping goals and colony size may vary, at the fundamental level, all beekeepers have similar objectives and must employ the same basic practices.

Novice beekeepers will have to address common startup issues such as finding a proper site for their new apiary, acquiring hive equipment and assembling it, sourcing bees to put in the new hive and doing everything possible to give them the best start in their new home.

With beekeeping comes great responsibilities, not only to the bees in your care, but also to your neighbors. Every beekeeper must do everything in their power to keep any disturbance from bees to a minimum, and to reassure and where possible educate their neighbors about bees.

By adopting an informed approach, what initially seems confusing and uncertain will become clear. Becoming a proficient beekeeper will take a few years, but they will be enjoyable years; as time passes, you can explore new aspects of beekeeping.

# 01 I don't know how to start beekeeping

## CAUSE

**Basic beekeeping requirements are not technologically difficult, but to the uninitiated the biological vocabulary and hive equipment nomenclature can be intimidating.**

## SOLUTION

In most aspects of beekeeping, techniques and management systems vary. This can be overwhelming and confusing for the novice beekeeper. Until you become familiar with fundamental beekeeping procedures, it is best to stick to the basics. Bees are very adaptable and if the issues of bee health, queen quality, and food availability are all adequately addressed, the colony will most likely be quite successful, regardless of what hive design or management procedure is used. Start by getting informed on the following areas:

- Where to find help and support (for example, beekeeping clubs and friends, academic programs).
- Where to purchase equipment.
- Where is best to locate your hives.
- Where and when to purchase live bees.

Books on beekeeping tend to offer too much detail and too many options, so start by reading concise and friendly educational materials to familiarize yourself with beekeeping terms and the basics listed above. Online video tutorials can also be a useful resource.

The best possible source of information, however, is experienced beekeepers, so find out if there are any established beekeepers in your local community, or even better, a beekeeping club. By speaking to them you can find out about the various styles of beehives and management procedures in common use in your area.

❋ *Information about beekeeping is widely available, both in print and online, but by far the best resource is other experienced beekeepers.*

# 02 I'm unsure when is best to start beekeeping

## CAUSE

Locating a bee supplier and purchasing proper beginning equipment will take time in itself; then comes the question of what time of year is best to set up your beekeeping venture.

## SOLUTION

In temperate climates, new bee colonies should be started during the warm season of the year. To survive the upcoming winter, the new colony must have time to gather and process pollen in order to build up its honey stores and population. Another option is to purchase an already established colony. These can be purchased and relocated at any time of the year.

You can start your research at any time of the year; however, beginning the decision-making phase during winter months would probably be the most practical, as this allows lead time for becoming acquainted with the craft, ordering and assembling equipment, and finding a source for live bees before spring arrives. If you wait until the spring months to begin this planning and research stage, it will most likely be too late in the year. The season will already be underway, leaving no time for ordering both equipment and bees.

A good time to make enquiries about a supply of bees is during late summer. To market their honey crops, beekeepers often have booths at farmers' markets. They are also a good source of information. If the beekeeper has been proficient enough to acquire a honey crop, he or she will be knowledgable about the craft and how to get started.

## POLLEN COLLECTORS

Forager bees are covered in a mat of soft fuzzy hair. Under a microscope, the hair is seen to be branched (plumose), whereas human hair is simply a single shaft. All three of the bee's body segments—head, thorax, and abdomen—have abundant hair on them. Even the bees' compound eyes have hair growing from them and the bee's head is covered, too. Pollen clings to this hair and is combed off by the grooming bees back at the hive.

❋ *Bees gathering pollen during the spring. In order to survive the winter, a colony must have time to gather and process enough pollen to build up the honey stores and the colony's population.*

# 03 I don't know where to get my first bees

## CAUSE

Most novice beekeepers are unsure how to go about procuring live bees. Reliable sources for bees and the related costs have increasingly become a challenge. Established beekeepers are also anxious to find replacement bees for winter losses and to increase their colony numbers.

## SOLUTION

If you are new to beekeeping, it is worth attending bee-related events and speaking to established beekeepers about the upcoming availability of spring season bees. You can find out about such events through local beekeeping publications and associations, and blogs and forums online. In the United States, package bees are a common way to start up a beekeeping operation. Bee packages, which usually contain three pounds of bees with a newly mated queen, must be ordered in the fall for subsequent spring delivery.

The spring season is the most intensive period for acquiring bees. During this time, colonies can be initiated from nucleus colonies (a small colony of varying size with a mated queen used as a "start" and normally called a "nuc") or other such splits from an established colony. Though rare now, established colonies sometimes become available when their owners retire from the business. You should expect to pay a high price for these producing colonies.

One of the primary challenges to the new colony and the new beekeeper is the first winter season. Bear in mind that, to survive the winter, any split, nuc, or established colony will need nectar that has been converted to honey, pollen that is stored as bee bread, young, healthy bees, and a productive queen.

✳ *A group of novice beekeepers with their instructor, inspecting a WBC bee hive at a local association apiary. By speaking to established beekeepers at bee-related events and beginner courses, you can find out about the best local sources for new bees.*

# 04

# I'm concerned that my apiary is in a poor site

## CAUSE

Apiaries are sometimes placed in less than ideal locations, either because the beekeeper had no alternative, because they made a poor decision, or because the location was intended to be temporary while hiving a swarm or dividing a colony took place, but it became an established site.

## SOLUTION

The ideal solution is to ensure that you select a suitable location for your apiary in the first place. When doing so, the following should be taken into account:

- In temperate climates, if water is consistently available, the hives will not suffer from being situated in the sun. In hot climates, some shade may be needed; manipulating the hives in full protective gear under the blazing sun can be hot work.
- Protection from cold winds lessens the demands on the colony during winter, helping to conserve honey stores.
- Consideration should always be shown to non-beekeeping neighbors, especially if they have children and pets.
- The apiary should be easily accessible to you, the beekeeper.
- It should be near to food sources for the bees.

However, if you have already set up your apiary and you find it to be lacking in the first three points above, modifications could help. Where sun exposure is a concern, ramadas or improvised shades can be erected around the hives; likewise, wind breaks can easily be added. Providing barriers such as fencing or shrubbery will force the bees to gain altitude as they leave the hive, taking them safely up and away from your neighbors. If the site is truly a poor one, the beehives may need to be relocated.

## RELOCATION, RELOCATION

Hives should only ever be moved incrementally. Abruptly moving a hive several feet (or farther) during daylight hours will result in the bees flying back to the original location. Alternatively, you could gently reposition the hives during the winter, when the colony is essentially dormant.

✳ *A commonly overlooked factor is the scenic setting. Most modern beekeepers keep bees for enjoyment and fulfillment rather than monetary gain. A scenic, placid apiary with gentle, natural sounds offers a quiet break from the typical workday schedule.*

# My new bees arrived during a cool, rainy period

## CAUSE

Bee package producers normally sell their entire inventory every season. To avoid disappointment, purchasers must place orders months in advance, but the weather on the delivery date may not turn out to be conducive to releasing bees.

## SOLUTION

In the rain—especially heavy rain—bees can be knocked from the air to the ground, where they will either starve or drown if they are unable to take flight in a few hours, making release during rainy weather inadvisable. If your package bees arrive on a rainy day, you can leave them in their package in a cool, dark room or basement for several days without undue harm until weather conditions improve. During this time, even though the packages have a sugar syrup supply within the cage, you should mist them with sugar syrup about twice a day. Watch for dead and dying bees on the bottom of the cage. A few hundred bees is normal, but a full mat of bees covering the cage floor about ¾in (2cm) deep is a cause for concern; if you see this, you shouldn't delay the bees' release any longer.

On nice days, package bees are commonly released by simply opening the cage and shaking the bees out. If no colonies are nearby, the bees will ultimately find the hive. However, during cool weather, the package should be released slowly to prevent too many bees from taking flight. One common technique is to place an empty hive box on top of the hive body that contains frames. The package is gently opened, the queen cage positioned, and the cage laid on its side within the empty deep hive body. Without undue confusion, over the next few hours, the bees will slowly leave the package and surround the caged queen.

# 06 Beekeeping equipment is expensive

## CAUSE

**Whether you're a novice beekeeper looking to set up your first apiary, or an established beekeeper in search of additional or replacement equipment, the cost of sourcing new materials can be very offputting.**

## SOLUTION

One way to minimize costs is to purchase secondhand equipment. If taking this route, it is important to first of all evaluate the physical condition of the equipment on sale: for example, look for any signs of rot, and find out whether or not the equipment was commercially manufactured. If the equipment needs repairs, these should not be extensive. Since they are inexpensive and prone to failure over time, wooden frames and combs are rarely worth recovering.

Secondly, if you already have existing equipment, will the new items blend in and be readily usable with your current setup?

The third consideration is the selling price. Having a beekeeping equipment catalog at the sale would be very handy for price estimations. How badly do you want this preowned inventory? The purchaser should have an idea of the cost of new equipment and the amount of time needed to assemble new equipment.

Fourthly, and the most important, as best you can, determine if the equipment is free of American foulbrood. Ask the seller why there are no bees in the hive. If you are suspicious, do not purchase.

If the equipment looks good, is disease-free, needs few repairs and is in the style of any equipment you are already using, and if the purchased inventory will augment your operation, make an offer near the selling price. If this is an impulse buy, only purchase it if is a true bargain. In the end, be prepared just to walk away and search again.

# 07 I'm not sure how to buy an established colony

## CAUSE

Buying an established hive of bees can be a confusing process because there are few standardized aspects to it. The size and health of the colony, the condition of the equipment, and the season are all factors that must be considered.

## SOLUTION

Purchasing an established colony is much like purchasing used equipment, but a live bee nest is included. If you are interested in the beehives primarily for the bee population, then a thorough internal examination is in order. Many states in the United States have apiary inspection services that hold disease inspection records. Varroa and other diseases and pest infestations should make the purchaser wary, but above all American foulbrood (AFB) is the key issue to be on the alert for when buying and selling bees. Be familiar with the appearance of AFB (see Problem 69) or ask someone who is acquainted with the disease to go along with you when shopping for an established hive.

In spring, a good beehive will cost more than it would in late summer. This is because with the difficult winter season already passed, and if headed by a prolific queen, a surplus of honey is logically expected.

The condition and quality of the woodenware is, of course, an issue, but it not as important as the health of the colony. A surplus honey crop can be produced or multiple splits can be taken from the strong colony. That potential makes the value of a few bee boxes and related equipment a minor concern. But everything is a variable. If you don't already own equipment, the state of the hive components will be more of a consideration for you.

## A RARE FIND

In recent years, due to the popularity of beekeeping and the scarcity of healthy bees, established hive costs have increased significantly. Unless a beekeeper is retiring or has experienced health issues, it is now uncommon to see productive colonies for sale. Expect to pay a high price.

✳ *These hives were put on the market by a retiring beekeeper. The colonies were strong and had been kept in a healthy condition, but the equipment was a mix of commercial and home-produced. The honey crop contained in the supers added to the selling price.*

# 08

# My nextdoor neighbors are afraid of bees

## CAUSE

**Stinging responses are a well-known attribute of most bees. Many people fear their reactions to stings will be deadly. In fact, this rarely happens, but their instinctual fear remains, even though they realize that bees are necessary for much of our food production.**

## SOLUTION

You should always strive to be a good beekeeping neighbor; keep your bees under control and do not try to manage too many hives. Beehives should be kept well away from boundaries and common passageways. If applicable, all zoning regulations concerning beehive locations should be followed explicitly. If hives are located on a small plot that puts the hive near to the neighbor's property line, you should install fencing or fast-growing shrubbery approximately 6ft (1.8m) high, which will force the bees to gain altitude quickly upon leaving the hive. Placing hives around buildings will accomplish the same objective. To discourage bees from visiting nearby water sources, such as your neighbor's pond or birdbath, a dependable, easily accessible water supply should be made available to them. Hives should primarily be manipulated at times when the neighbors are not nearby. Bee swarms, though usually gentle, are particularly frightening to inexperienced people. If at all possible, prevent swarms from departing and moving to neighboring property.

If you are doing all of the above, you will be in a good position to reassure the fearful neighbor. Never boast about stinging experiences, and when the honey crop is processed, regularly offer them the traditional honey gift. It is not just the neighbor's concerns that should be addressed; remember to reassure any tradesmen or other visitors to yours or your neighbor's property.

## PROTEIN HUNGRY

In spring months, protein-starved bees will aggressively seek out bird feeders and animal feed bunkers. This can frustrate both the neighbors and birds and other animals that must deal with bees at their feeders. Before the season starts, offer your own protein substitutes to your colonies and let the neighbor know that the bees' stay at the feeders will quickly pass. Such foragers are not aggressive, but can cause concern.

❋ *These British hives are situated next to a high wall, which will force the bees to fly upward on leaving the hives, keeping well out of the way of the neighbors.*

# 09 Some bees were crushed during a hive examination

## CAUSE

Modern beekeeping practices can result in a much larger population of bees than would normally occur in a natural nest. During active times of the year, bees will appear to boil from the hive. Repositioning frames and resetting hive bodies always results in some crushed bees.

## SOLUTION

There is no practical solution that prevents all bees from being crushed or injured during a hive manipulation. Commercial beekeepers with many colonies to manipulate rarely have the luxury of quietly and gently working colonies. Regardless, all beekeepers should make all practical efforts to keep harm to a minimum. Aside from the undesirable loss of life when bees are crushed, pheromones are released that could cause the colony to be become agitated.

You can use smoke to push bees back into the colony or to send them in another direction. Heavy smoking isn't usually required, except in a case of a populous colony, where light smoking will doubtlessly move a few bees, but not enough and not quickly enough. Keeping a large colony open too long will increase the risk of stinging and loss of control.

Brushes can be useful to a limited extent. A brush can be used to physically move bees out of the way, but bees are quickly made unhappy by being brushed around—especially with a brush made of animal hair.

As soon as you are experienced enough to work a colony without gloves, your manipulations and movements will become more precise and controlled.

## OPERATION CLEANUP

The colony has no way to clear dead and injured bees other than either dragging the bee corpses from the hive or ingesting them. The bees will make every effort to keep the healthy colony in pristine condition, so leaving crushed bees in place is not an option. One way or another, the dead bees and bee parts must be removed.

✳ *A colony of this strength cannot be opened, reassembled, and closed without some bees being in the wrong place. Save as many bees as possible by using smoke and a brush when manipulating the hive.*

# 10 My bees are aggressive toward garden machinery

## CAUSE

**Bees can become irritated when lawn mowing or gardening equipment is operated near the colonies, but it is usually because other factors have already caused them to go on the defensive.**

## SOLUTION

Bees can become defensive for a number of reasons: scarcity of food sources, the colony being at full population, and disturbance from wildlife, to name a few. Weather changes can also have an effect on the personality of a colony: for example, rainy weather could have forced an increased number of older bees to be at the hive. Once the bees are in this agitated state, they will be in no mood for putting up with a lawn mower nearby. Gasoline-powered equipment seems to be particularly offensive to bees.

When operating string trimmers, mowers, or garden-tilling equipment, particular care must be exercised. If working near the colonies, at the very least a veil should be worn and a lighted smoker should be periodically used. This would provide an escape strategy if one is needed. Raising the hives higher from the ground would provide increased mowing space. Rather than use noisy equipment, some beekeepers judiciously apply herbicides to keep hive entrances clear of vegetation.

Always keep bystanders in mind. While stinging attacks are rarely noteworthy, to the uninitiated who have no protective gear, such events can be frightening or annoying.

❋ *The beekeeper should always be protectively suited when operating garden machinery near the bee colonies, in case it provokes a hostile reaction from already defensive bees.*

# CHAPTER TWO
# BEEKEEPING EQUIPMENT

To the uninitiated, beehive equipment and protective clothing can appear complicated. This view is deceptive. The only purpose of the protective gear is to shield the beekeeper from an excessive number of stings; at the most basic level, the beehive is little more than a stack of boxes, each holding about eight to ten framed honeycombs. Everything else is personal choice.

All bees really need in order to set up a colony is a dry, dark cavity approximately one cubic foot square, with a defendable entrance and nothing else already living there. Consequently, bees are amenable to most common hive designs.

Confusion can arise due to the sheer number of designs and styles of hive in use—from the popular Langstroth to the British National and WBC designs, and from Warre to top-bar hives. Some of these styles have been pictured in the book to give a feel for the variety available around the world, but the best advice is always to find out what equipment is most commonly used in your local area. Though daunting at first, with the help of appropriate study materials and possibly a beekeeper friend, all will become clear. Once the initial choices have been made, the novice beekeeper can begin developing expertise, and natural expansion will occur.

# 11 My hive equipment doesn't match

## CAUSE

There are many hive designs available around the world, ranging from simple to surprisingly complex. Occasionally adding to the confusion is the fact that manufacturers' measurements and designs may differ slightly. Mixed equipment may fit together, but not perfectly.

## SOLUTION

Since bees are not particular about the appearance of the hive, to a great extent it is not a problem if equipment doesn't match. Different hive types can be improvised to work together in many instances. Nearly any beehive composed of vertical, free-hanging frames can be modified to incorporate different styles of equipment. However, there is a point of diminishing returns. Beehives are already a conglomerate of parts and pieces and to add even more nonstandard parts and pieces only creates further confusion. In springtime, reversing brood boxes may become difficult and would probably require individual frames to be shifted rather than entire boxes. And while the bees may not care about the outward appearance of their hive, the proud beekeeper may derive a certain amount of satisfaction from tending a neat, well-turned-out apiary.

When setting up a new beekeeping operation, the best advice is to opt for common hive designs that are used and available locally. That way, when the time comes that you need additional or replacement equipment, it can likely be acquired from a supportive beekeeper in the area. Additionally, when purchasing used equipment at sales and auctions, there will be a greater chance that the equipment on the block will be compatible with the equipment you already own.

✳ *The three styles of United States equipment shown here—8-frame, 10-frame and expanded polystyrene—are commonly available, but individual components will not fit together without significant modifications.*

# HIVE SWEET HIVE

Bees do not have a "favorite" hive design. At different times, they both thrive and fail in all types of equipment. The hive design is primarily a decision made by the beekeeper. Under normal conditions, the bees are generally agreeable with this beekeeper decision, but if it is not suitable, it is not uncommon for a colony to abandon a hive.

# 12 I'm concerned about hive equipment theft

## CAUSE

In nearly all parts of the world keeping bees is presently a very popular undertaking. Consequently, honey bees, in hives or otherwise, are in short supply and the value of equipment has increased significantly. Often, apiaries are located in areas that allow easy access for opportunistic thieves.

## SOLUTION

The obvious security measures, such as installing electric fencing, an alarm system, or a locked enclosure, may well prove too expensive for the small-scale beekeeper. Marking your hive equipment can act as a deterrent to thieves and is much more readily affordable. It also provides proof of ownership in the event that any stolen equipment is tracked down.

Branding wooden equipment has historically been the most common method of marking wooden equipment. Equipment should be marked before painting with an obvious brand mark. Buying a large branding iron may not be an option for beekeepers who have only a few hives to mark—you could look into borrowing the equipment. Your local beekeeping club may be able to help. Though considerably less expensive, smaller branding irons make marks that are more easily obliterated.

Hives can be painted characteristic colors, but obviously they can be repainted. An identifying mark can be put inside hive boxes, but that will require access to the equipment in order to identify it.

Presently, technology is being developed for hives similar to that used to find lost or stolen mobile phones. However, this technology is nascent and may not be affordable for most beekeepers. For now, branding wooden equipment is the most common hive-marking procedure.

❃ *These hives in Italy are distinctive thanks to their brightly colored paint, but an opportunistic thief could easily paint them a different color. A brand mark would be much more difficult to cover up.*

# 13 The paint finish on my hives is failing

## CAUSE

If hive bodies are painted at all, the inside surface is rarely coated with paint. Consequently, moisture from the high humidity hive interior wicks through the wood and related joints to cause the paint finish to fail from the inside—not necessarily the outside.

## SOLUTION

The average hive body, in constant use, generally lasts about seven years. There are several possible methods to help the hive bodies and supers last longer and make them look neater. The easiest procedure is not to paint the woodenware at all. It will take on a weathered, rugged look that is not unappealing. Interestingly, over the years, bees will coat the inside of the unpainted wood with propolis and wax and this will coax more time from the hive boxes than would be expected.

The second method is to paint the equipment—inside and out—at regular intervals of about three to four years. Commercial beekeepers will sometimes spray paint large quantities of stacked equipment at once, especially if the equipment is used for honey production.

Thirdly, you could use an exterior stain product with an UV inhibitor. This is normally a product used on the exterior surface of houses and cottages. Such finishes are particularly popular in Canada. This gives a natural and appealing color to the equipment but abundant time should be allowed for the finish to cure, not just to dry.

In any case, painting the edges is a troublesome task. So troublesome, in fact, that many beekeepers forego the effort. While paint does help protect the exterior wood surface, primarily the equipment is being painted for esthetic reasons.

## UNSIGHTLY MOLD

When hive boxes that are used for honey production are stored, the slight film of honey on the surfaces can support mold growth. It is not a pathogenic mold, but to many people it is unsightly. A pressure washer is useful to keep the equipment clean and ready for a fresh coat of paint. Bees don't have any preference whether equipment is painted or not.

❋ *The hive shown on the left has been painted with an exterior penetrating stain normally used to coat lake cottages and cabins. It will need to be recoated about as often as regular paint, but the finish is visually appealing. The hive on the right has been painted with exterior latex paint. As is expected, after a few years of use, the finish begins to age and mildew growth starts.*

# 14 The inner cover of the hive is glued down

## CAUSE

Either the colony has become very crowded or the inner cover was incorrectly left in the inverted winter position. In either instance, the bees have built heavy bands of brace or ladder combs between the top bars and the inner cover surface, making it difficult to open the hive.

## SOLUTION

This issue is a common occurrence. This can be seen in the general design and fitting of the inner cover. From the outside, the joint between the top edge of the super and the lower edge of the inner cover is readily accessible with a hive tool. If a colony is seriously crowded—even if the equipment manufacturer respected bee space measurements—combs containing either honey or drones will be jammed into all available small spaces. Indeed, inner covers on hives that are not opened for several seasons will become so tightly stuck that they may actually break apart when forced off the hive. Just beneath the inner cover, frames, both wood and plastic, will also be soundly stuck to the hive body. In general, this hive will be difficult and messy to manipulate.

The only solution is to regularly scrape off propolis and brace combs whenever the colony is opened. An alternative would be only to remove selected frames and to keep those frames reasonably clear of propolis and extraneous combs. To a degree, always providing ample hive space would mitigate the behavior. The bees will diligently work to glue and seal all components back into place almost as soon as the beekeeper completes the task of removing these natural materials. This activity primarily occurs when plant resins are available or when nectar flows are ongoing.

## FILLING IN THE SPACES

If constructing a nest in an enclosed cavity, a bee cluster has the tendency to initiate nest construction at the highest point. The bridging behavior shown in some colonies may be a response to combs not being soundly attached at the top. Filling in space between frame top bars, bottom bars, and the inner cover will approximate a more natural nest design.

❋ *The primary function of the inner cover is to prevent the rimmed outer cover from being glued to the top of the hive. The inner cover can then be pried from the hive top using a hive tool at the joint formed between. The one pictured is in the inverted winter position.*

# 15 Plastic hives accumulate water during the winter

## CAUSE

Nearly all artificial hives will accumulate excess moisture under certain conditions. Since they are nearly impenetrable to water, unventilated expanded polystyrene hives will be particularly troublesome in this regard.

## SOLUTION

During winter months in temperate climates, most beehive styles, whether made of wood or plastic, will accumulate moisture. Some moisture is a requirement within the healthy hive: the humidity within the brood nest needs to be at least 60 percent. Yet, possibly due to hive design, a significant amount of water can accumulate within the hive, which is caused by the metabolic activities of the wintering bees. Warm air rises from the wintering cluster and when it hits the top of the hive, the air cools and ice forms. Over the winter season, quite a bit of ice can collect. This is essentially no problem until spring, when the ice melts and drips cold water on the cluster below.

The natural nest has some extraordinary techniques for dealing with moisture. In beehives, the primary solution is to ventilate the hive near the top, just as houses are ventilated. This allows the moisture-laden air to escape without forming much ice.

Plastic hives, being essentially impenetrable to water, can accumulate a surprising amount. This was primarily an issue with early styles of plastic-hive equipment. Even with several inches of water in the hive bottom, the cluster seems to winter very well. Adding a screened bottom board to a plastic hive allows water drainage and improves air flow, while also allowing Varroa mites to fall out of the colony.

✿ *This early styled plastic hive was rugged and well made. The warm wintering cluster produced a good deal of water that accumulated over the winter months. Adding ventilation and anticipating the moisture issue resulted in a productive colony the next spring.*

## PLASTIC FANTASTIC

Due to their insulating characteristics, dense, expanded polystyrene beehives have become a commonly accepted hive style. Water accumulation is an issue, but it is one that has been readily addressed in the most current hive designs. Beekeepers hope that the wintering colony will be better insulated, allowing it to withstand the rigors of a long winter.

# 16 Beeswax foundation is difficult to install

## CAUSE

The greatest issue when installing beeswax foundation is whether the foundation is the proper size for the type of frame being used. Foundation inserts can be inflexible, making them difficult to install.

## SOLUTION

If your foundation sheets are too long to fit within the inside boundaries of the frames, and if only a few sheets are to be inserted, they can be cut down with large scissors or snips. If they are too short, leaving a space along the bottom edge, bees will tend to fill in the gap when constructing combs. For a good fit, the most logical solution is to purchase the frames from the same company that manufactures the foundation.

You must also decide what type of foundation to use. The earliest forms of bee comb foundation were made from 100 percent beeswax but this product was weak and would sag in hot weather. Later, thin plastic sheets coated with beeswax and embossed with cell base outlines were added to provide strength. Rather than sag, if not supported at the ends, this improved hive product would bow. Though all of these foundations are still being produced, currently the most popular design is plastic frame inserts. The frame insert is a thick beeswax-coated plastic sheet that has been embossed with cell template outlines. It does not sag or bow and requires no special frame insertion techniques or tools.

The foundation insert can be flexed and snapped into place within the frame. Occasionally, the insert sheet will bind within the frame, making it difficult to reposition the sheet. Alternatively, the foundation can be positioned inside the frame before the bottom bar is in place. The bottom bar is then used to trap the insert in place without having to flex and snap the insert sheet.

✳ *Snapping foundation inserts into new frames is a simple process. First, the frame is completely constructed. Then, the foundation insert is pushed into grooves that are cut into both the top and bottom bars of the assembled frame. The insert will snap into place and become trapped in the grooves within the center of the frame.*

## WIRED FRAMES

When using beeswax foundation, "wiring" the frames can help support the combs when they are being extracted. The wired-frame procedure is now considered to be a classic aspect of beekeeping. The amount of assembly work involved in preparing the frame is considerable, but it will result in reinforced wax combs. Some beekeepers enjoy the ambience and formality of the technique.

# 17 I don't have the right tool to open up the hive

## CAUSE

Every beekeeper should be equipped with a purpose-made hive tool for prying open the hive. But it is easy to misplace these necessary implements, or drop them when a beehive manipulation is under way.

## SOLUTION

As emergency need dictates, nearly any type of pry bar can be used as an improvised hive tool. But over time, the use of such improvised tools will damage the frames and edges of the hive bodies, shortening the equipment's useful life. Additionally, they are not as efficient as traditional hive tools, making manipulations lengthier and increasing the risk of inflicting damage within the hive. So, it really does make sense to ensure you have the correct tool for the job. Experienced beekeepers will accumulate several of these valuable tools—it is always a good idea to have a spare on hand in case of accidents.

For many years, there were two kinds of hive tools—a long one and a short one (9½in or 7in/24cm or 18cm). Now, modern bee supply catalogs offer multiple versions of what at one time was little more than a "window opener." These implements are made from hardened spring steel and will tolerate a good deal of abuse. Any of the versions now offered work very well for removing stuck frames and for separating hive body equipment.

If a frame is solidly stuck with propolis and wax, the frame top bar may either pull from the end bars or break under heavy prying pressure. For frames stuck this badly, stand the hive body on end and from the bottom tap at the frame ends with a hammer, driving a frame or two from the hive body. Once one frame is out, the remaining stuck frames can be readily removed.

## HIVE TOOL PITFALLS

An unfortunate mistake that can be made with a hive tool is to slip it in your hip pocket, forget that you put it there, and get into your car. There is an excellent chance the car seat will be damaged. When using a hive tool to pry open soundly stuck hive equipment, a glued joint can abruptly break apart, resulting in jammed thumbnails.

❊ *Around the world, hive tools are manufactured in many styles. Features common to all are wide blades on both ends for prying frames and scraping propolis.*

# 18 The queen excluder is disrupting colony function

## CAUSE

Many beekeepers believe that queen excluders restrict free worker bee movement into the supers, thereby causing honey to be stored just beneath the excluder or, alternatively, no honey to be stored at all.

## SOLUTION

Interestingly, this bee management problem is not viewed as a troublesome issue by all beekeepers. The basic function of the queen excluder is to restrict the movement of the queen into honey supers. The grid, having openings of $\frac{5}{32}$in (0.4cm), are made of metal or plastic. This grid allows workers to pass but restricts queens and drones from moving into honey supers. This measurement is critical, requiring excluders to be handled carefully; if there is any damage, the queen will find the resulting opening and move through.

The argument against using excluders is that nectar-loaded bees are reluctant to squeeze through the narrow openings and that the grid causes undue restrictions on the colony. This concern will seemingly not be resolved in the near future. The excluder is an established piece of bee equipment with both beekeeper supporters and detractors. When there are many colonies to manipulate, a beekeeper will probably use them, while beekeepers with fewer colonies can spend the time to be certain that the queen is not with brood in a honey super that is removed for extracting. There are different styles of excluder, with pricing based on quality of manufacture. Plastic excluders are the least expensive.

It comes down to personal choice. If queen excluders don't fit in with your colony management philosophy, don't use them. But bear in mind that you will have to deal with the occasional spot of brood in honey supers. The true risk of not using an excluder is that the queen is unintentionally removed from the hive along with a honey super.

❋ *A beekeeper at work removing wax from a zinc queen excluder. Excluders must be handled carefully to ensure no damage as the queen will easily find any resulting holes. Excluders may be made from various materials; some beekeepers prefer to use plastic.*

# 19 My bee gloves make hive work difficult

## CAUSE

**When working beehives in hot weather, or if gloves become honey-soaked, finesse and manual dexterity become an issue.**

## SOLUTION

Honey bees can become defensive if their colony is threatened. In extreme cases, such as an accident where large numbers of hives are spilled and great numbers of bees are confused and in flight, full protective clothing is required. A veil, heavy coveralls, boots, and heavy gloves with elastic cuffs make up the necessary protective clothing. In routine hive examinations, the veil is normally the only necessary equipment. Typically, experienced beekeepers do not wear gloves during general colony examinations.

If you are new to beekeeping, there is no shame in wearing full protective gear, including gloves, until you build up your confidence and experience. Beekeepers with many hives to manipulate may also choose to wear them, so that they can work quickly and efficiently.

Every type of glove has its shortcomings, and they all add a degree of clumsiness to hive manipulation. Plastic and rubber gloves hold water generated from perspiration. Lightweight gloves such as those used for medical procedures are too thin and bee stingers can penetrate them—you could try wearing two pairs, which would allow you to replace the outer pair as you move between hives, thus reducing the risk of transferring disease. Heavy canvas gloves are more inexpensive and will work for a while. Leather or goatskin gloves have a much longer useable life, but they are almost impossible to clean.

At the fingernail, a typical finger is ⅜in (1cm) thick; with gloves on, this grows to about ½in (1.3cm) thick. Bee space between frames averages ¼–⅜in (0.6–1cm). Getting oversized gloved fingers between frames will prove difficult, so the sooner you can move on to working without gloves, the better.

✳ *The gloves pictured are made from leather with rigid, extended cuffs and have ventilated wristbands. An elastic band prevents the glove cuffs from sliding down the arm. You should always clean gloves and tools in a solution of caustic soda after use.*

## SUPPLE FINGERS

The fingers on beekeeping gloves frequently wear through, exposing the vulnerable fingers inside. Many beekeepers use tape over the worn areas to squeeze a bit more service out of the gloves. However, the tape just adds to clumsiness. If possible, acquire a high-quality pair of ventilated gloves. Occasionally soaking them in boot oil will help to keep them pliable.

# 20 The beehive smoker will not stay lit

## CAUSE

**The air needed to keep the embers inside a smoker lit is supplied by the bellows. If the bellows are not flexed every few minutes, the embers will start to go out.**

## SOLUTION

Beehive smokers are necessary devices for successful hive manipulations. The smoke they produce temporarily masks the internal chemical communication system of the colony. Consequently, bees are unable to assemble a coordinated defensive response, and the beekeeper can manipulate the hive with reasonable safety.

The effects of smoke vary, but in general a beekeeper can expect about ten minutes of protected work time before the effects of the smoke wanes. The problem arises when the smoker bellows are not flexed for ten minutes or so. The embers producing the smoke cool due to a decreased oxygen supply. As the bees begin to reassemble their defenses, the beekeeper reaches for the smoker—and when needed the most, it has died out. With the hive opened and bees becoming increasingly agitated, the entire lighting process must be performed. Take heart. All beekeepers find themselves in this scenario at some time.

To reduce the risk of this happening, be sure to build a good ember bed in the smoker at the outset. Start with newspaper or pine needles. Pump the bellows until an open flame appears. Slowly add more fuel and bring back to a full flame. Continue the cycle of adding fuel and bringing to a flame. Once smoke is easily produced, close the lid and frequently pump the bellows. You are looking to produce copious amounts of cool, white smoke; bluish, hot smoke agitates bees, can sear their wings, and should not be used.

## FUEL FOR THE FIRE

There are various fuels that can be used in smokers. For a quick hive examination, use fast-burning fuels such as pine needles or dried leaves. For a longer burning fuel, wood shavings, burlap cloth, rotted wood, or cloth rags are commonly used. For your own good and the good of the bees, in all instances use the least amount of smoke possible.

❋ *Gentle smoke whiffs from a quiet smoker are an indication that the embers are still hot enough to combust. A large smoker, properly stoked and periodically recharged, will provide smoke all day. Over time, as the ember bed grows, the smoke will become very hot. The wire grid is to help keep fingers and hands away from the hot barrel. Smokers can cause accidental fires so appropriate care must be taken.*

# 21 The handles on my bee boxes are inadequate

## CAUSE

Generally, the thickness of wood used to manufacture beehives is ¾in thick. Consequently, the standard recessed handle that is cut into the sides and ends of the boxes can only be about ⅜–½in (1–1.3cm) thick. If gloves are used, this does not provide for much of a handle.

## SOLUTION

At most, hive body handles are ¾in thick and most likely they are considerably more shallow than that. Additionally, honey is remarkably heavy—and guarded by great numbers of potentially angry bees. Beekeepers often wear bulky protective clothing, and the ground around the hives is not always level and solid. Lastly, nearly all beehive styles result in a box that is clumsy and heavy to carry. Obviously, conveniently positioned handles would seem to be completely logical, yet an entirely satisfactory solution to this problem is yet to be found. But there are a couple of things you could try.

An alternative to the usual recessed handles is to attach a wooden strip along the top of the recessed handhold of the hive box ends to provide more of a grasping point at each end of the box. These strips should be firmly attached to the hive body ends with cinched nails, or screwed in place. If the box is unpainted, gluing will be very helpful.

If this doesn't prove helpful, you could consider using shallower boxes. These will still have the same restrictive handle issues, but the reduced weight makes them easier to carry.

✳ *The recessed handles on this hive box allow the beekeeper to pick up the box with ease.*

# 22 A hand truck loaded with a colony tipped over

## CAUSE

**When a hand truck is heavily loaded with beehives or full honey supers, it can become unstable and difficult to maneuver on the soft ground of an apiary.**

## SOLUTION

Beekeepers commonly use carts, wagons, or hand trucks to move equipment around both in the yard and in the storage house. Such carts are manufactured in many different sizes, styles, and price ranges. In general, they are useful to have at the ready for moving hives, supers, or equipment that is in storage. Problems can arise when two-wheeled hand trucks are used to move heavy hives or full honey supers in the apiary. The wheelbase is so narrow that a top-heavy hive will cause the cart to become unstable, especially on soft ground. If a live colony is being moved and the load tips, it will result in many defensive bees spilling from the hive, and it will be difficult to regain control.

Common sense is the best solution when it comes to maneuvering a loaded hand truck. When possible, move the truck along firm paths. It is also a good idea to ratchet-strap the equipment to the cart. Pulling the loaded cart is a better option than trying to push it on soft or uncertain ground, and ask someone to help you.

Though most bee supply companies list hand trucks in their catalogs, there are no models that are specifically manufactured for the bee industry. In general, heavy-duty models are better suited, but they will also require more effort to get on and off the truck. Have a hand truck available, use it when necessary, but expect occasional problems.

❊ *The unit on the left is inexpensive and lightweight, while the larger device on the right is heavy duty and can withstand abuse and heavy weights. Two people will be needed to propel the heavy-duty cart in the soft ground of the apiary. Even so, this is less strenuous than carrying the full colonies or full honey supers.*

# 23 My observation hive is not thriving

## CAUSE

Observation hives are not normal domiciles for bees. Left to their own decisions, scout bees will search out dark cavities rather than one as openly lit as an observation hive. Nearly all of these units are high maintenance and require frequent manipulation, but an observation hive is always a popular educational exhibit.

## SOLUTION

In the natural state, a bee nest exists in total darkness so a colony will be less than comfortable in a brightly lit, glass-walled hive. If possible, the glass walls should be covered when not being viewed.

One- or two-frame units are only suitable for a short time. Such small hives are perfect for presentations and other short-term educational exhibits. If the colony is expected to exist in a glass-walled hive on a semipermanent basis, multiple frames help the colony develop a functional brood nest. Nine deep frames (3 on 3 on 3) are often enough to help the colony get through the warm season, but such a configuration will not always highlight the queen's activities. In fact, much of the time, she will avoid the outside frames, but for most observers, being able to watch so many bees safely behind the glass presents sufficient intrigue.

Be prepared to support and maintain an observation hive throughout the entire season, and to deal with swarms from colonies that become crowded and cannot be given more space. Most units will not survive the winter; you will more than likely have to disassemble the unit come fall.

An observation colony being viewed by the public must be kept in prime condition. A dead or dying observation hive will elicit concern and sympathy for bees that are obviously in a dire condition.

## FEEDING TIME

An observation hive should ideally be fed both sugar syrup and a supplemental protein source. While feeding carbohydrate is fairly simple, feeding the protein material is difficult. If the design allows the practice, feed protein. Otherwise, be prepared for a few bees to escape each time the sugar syrup jar is removed from the hive. Escaped bees will normally move to the nearest window.

❋ *The colony in this observation hive is not thriving. In fact, it is close to starvation. It requires emergency feeding of both carbohydrates and protein. Additionally, more adult bees should be added. Due to the greenhouse effect, an observation hive should never be left out in direct sunlight.*

# 24 I'm confused by differing foundation cell sizes

## CAUSE

From the earliest years of bee foundation manufacture, some controversy has centered on the best cell size to stamp on the wax foundation sheet. For the new beekeeper, this management topic can be particularly confusing.

## SOLUTION

New beekeepers are subjected to a significant number of initial decisions concerning things like hive types, frame styles, queen stocks, and methods of starting a new beekeeping project. Which cell size to opt for on the foundation is also an issue that comes up.

Much of the confusion surrounding this topic stems from the fact that cell dimensions tend to vary even within the same colony. Generally, in the United States, the discussion has centered around two sizes of foundation comb cell sizes: the ³⁄₁₆in (0.5cm) diameter cell, and the more common industry "standard" ⁷⁄₃₂in (0.6cm) cell. In 1933, Prof. Ursmar Baudoux began a discussion in the beekeeping literature that promoted larger cell sizes in order to produce larger bees that would, theoretically, produce more honey. In recent years, others have taken interest in this topic area, but from a different view. It was proposed that smaller cell sizes would produce bees with a shorter development time, which would disrupt Varroa development. While many beekeepers have anecdotally supported this observation, scientific studies have not clearly confirmed this effect.

New beekeepers, having limited to no beekeeping experience, should not become embroiled in this technical controversy until they have a good basis in beekeeping management and beekeeping experience.

# 25 I'm unsure which bottom board design to opt for

## CAUSE

**Until the screened bottom board became common, all bottom boards were made of solid wood. Now some bottom boards have screened parts while other bottom boards are solid.**

## SOLUTION

The bottom board style will be the beekeeper's decision. In a natural wild nest there is no bottom board. The style of hive the beekeeper is using will likely dictate which type bottom board can be used.

Presently, most hive designs offer a bottom board with a screened section. The screening is commonly 8-mesh (8 squares per inch/2.5cm). The reason for this screened section is that the occasional Varroa mite (see Problem 62) that is accidentally knocked or groomed from a bee will drop completely out of the hive. If the bottom board is solid, the mite will (theoretically) tumble to the bottom of the hive and land on the bottom board. It will await a passing bee and be carried back up into the brood nest area. Unfortunately, if used as the only method of Varroa control, the screened bottom board is not effective. No doubt it helps, but not enough to warrant not using other methods to reduce Varroa populations. Many of the screened bottom board designs come with inserts that allow the screening to be closed in winter months or when assaying the colony to estimate Varroa counts.

Beekeepers who frequently move their colonies for pollination or wintering purposes may not use screened bottom boards. Inserting the hand truck toe-plate under a heavy hive can cause rips or holes in the bottom board screening. Commercial beekeepers commonly use bottom board pallets that are designed for use with skid-steer loaders. Screened bottom boards are best used on hives that are rarely moved.

# 26 A loaded hive stand has accidentally collapsed

## CAUSE

**Beehives become surprisingly heavy as a productive season progresses. A hive stand may appear to be level and stable, only to shift or collapse when a nucleus colony that weighed about 40 pounds (18kg) grew to weigh around 300 pounds (136kg). At some point, the load is too great for the stand.**

## SOLUTION

Throughout the development of beekeeping equipment, a perfect hive stand has never been designed. It would appear that nearly any supportive structure could conceivably have a beehive set upon it. The challenge is that the hive stand needs to be just the right height and allow for a gentle slope to the front so rainwater will not accumulate within the hive. Tall stands are convenient for working small colonies, but a tall hive stand makes it very difficult to add and remove honey storage boxes. For colonies housed in hives that are frequently moved, a hive stand becomes yet another item to be loaded, moved, and then unloaded. It is not surprising, then, that so many devices have been tried as a support base for the typical beehive.

Cement blocks and treated wood timbers are common hive supports but the typical block is about an inch too narrow and the timbers are only 4in above the ground. Bee supply companies have historically offered simple structures that raise the hive just a few inches above the ground and have a front sloped landing board to assist heavily laden bees returning to the hive. In choosing a hive stand, go for a design that is stable, strong enough to withstand several hundred pounds of hive weight, portable if hive moves are anticipated, rot-proof, of desirable height (probably no more than 24in/60cm) and with an appearance that is aesthetically pleasing.

✻ *Not only do hive supports occasionally collapse from the weight of heavy colonies, but occasionally mowing equipment or farm animals bump hives. The hive shown in the photo was struck by the mower deck on a small tractor, causing the hive to fall forward. The hive stand was constructed of heavy treated lumber 8in (20cm) wide but the stand was narrow (approximately 18in/45cm) and did not sufficiently support the hive.*

## MASS GRAVE

It should be noted that the hive entrance being near ground level is a manmade convention and not an entrance location that temperate bees would normally choose. Departing bees carrying a dead comrade cannot become airborne before crashing into the ground. The result is the typical compost pile that accumulates at the entrance of traditional ground level hive entrances. This decaying mass attracts skunks, raccoons, and other insects, causing secondary problems.

## CHAPTER THREE

# BIOLOGY AND BEHAVIOR OF THE COLONY

It is easy for the beekeeper to grow to see the bee colony as a domesticated animal, one requiring management and care. Over the years, hive equipment and management procedures have evolved that show the bee in an unnaturally tamed light. In fact, the bee has another natural life that beekeeper management techniques unintentionally suppress.

In reality, the honey bee continues to adapt to the demands made by the beekeeper and the equipment used by typical beekeepers. But withstanding this pressure, the honey bee is still a wild being. No aspects of bee studies make this plainer than a review of the biology and behavior of the honey bee. When the naturalness of events such as swarming, queen replacement, brood care, and wintering preparation are observed, honey bees show that they still understand their natural life. Competent beekeepers strive to understand these aspects of the bees' natural world and incorporate these behaviors in their management systems.

It is important that you gain a solid understanding of bee behavior and colony biology, so that you can distinguish between what's normal and what represents a problem.

# 27 My bees seem to sting more often than is normal

## CAUSE

If a colony is genetically disposed to defensive behavior, the bees will readily sting in an effort to defend the colony and its resources. However, bees may also increase stinging levels if animals have been pestering the colony at night, if it has been exposed to insecticide, or if the beekeeper has been clumsy when manipulating the colony.

## SOLUTION

From season to season, beekeepers who closely monitor their colonies come to know the current personality of the colony. If an otherwise docile colony becomes overly defensive, you can either attempt to decipher the cause and correct it, or choose a different day to manage the colony. Some aspects of excessive stinging response are beyond the beekeeper's ability to manage. For instance, thunderstorms seem to make bees testy. Also, bees will be feistier during periods of heat and when there is an absence of nectar. During these periods, if hive manipulation is required, you should wear full protective gear and use abundant amounts of smoke.

If a bee colony is so defensive that it is unenjoyable to manage and is a danger to the neighborhood, the queen can be replaced with one from a reputable queen breeder. Hopefully, the new queen's genetics, reflected in her worker offspring, will provide a bee that is more docile and manageable. Additionally, you can reduce sting occurrences by applying white, cool smoke to the colony before it is agitated and then reapplying smoke at regular intervals. As much as possible, you should avoid bumping and jarring the colony when manipulating it.

Importantly, if you get stung a time or two, try to keep from over-reacting by waving and swatting and possibly dropping frames, as this will only make matters worse by agitating the bees further. In general, expect some stings, but don't expect too many.

## IT'S ALL IN THE STINGER

The trademark of a honey bee sting is the small stinger mechanism that will remain embedded at the sting site. The stinger should gently be removed, although no great harm is caused if this is not done immediately. Fishhook-like barbs on the sting shaft make it difficult to dislodge. Most beekeepers develop a physiological tolerance to stings over time and show reduced effects of swelling and pain.

✳ *While these bees may appear aggressive, in fact they are not. The beekeeper is manipulating a gentle honey bee swarm. If the bees were attacking, many more would be stinging the beekeeper's protective clothing. Also, aggressive bees would not quietly be sitting on leaves.*

# 28 A swarm has clustered beneath the parent hive

## CAUSE

Occasionally, a swarm from either the parent colony or a neighboring colony will cluster in the small space between the hive and the ground. Rainy weather, the physical condition of the queen, or the emission of attractant hive odors from the screen bottom, could be the cause.

## SOLUTION

Typically, honey bee swarms occur nearly every spring season. Taking the old queen, approximately two to three pounds of departing bees will temporarily bivouac on something like a tree limb for a few days while the final new nesting site is chosen. At this brief interval, this bee swarm is easy to retrieve and to install into hive equipment. While some swarms are easy to hive, others are much more difficult.

When the departing swarm clusters beneath the colony from which it issued or underneath a neighboring hive, it makes for a troublesome swarm retrieving experience. While this situation does not happen every season, it happens often enough to be an inconvenience. Hives with solid wooden bottom boards rarely have these swarms beneath them. Normally, temperate honey bees do not have much interest in being so near the ground. If the swarm is small, it can actually go unnoticed.

To retrieve the swarm, the colony must be completely broken down to the bottom board. It can be a laborious process. Additionally, some confusion can result when the swarm queens in the parent colony are disrupted and the swarm and the original parent colony bees are mixed in flight. The bottom board containing the swarm can then be carried to an awaiting empty hive filled with combs. The size of the hive box is irrelevant at this time. The bees can be shaken in the new box or the swarm can be laid in front, allowing the bees to find the cavity themselves.

✳ *This photo was taken from the rear of the colony and the back edge of the bottom board, with the closing panel removed. The swarm is only a few inches from the ground. It has no future in this location and should be removed by the beekeeper.*

## SWARM BEHAVIOR

Though common, and though swarms have similar biological attributes, there is no "standard" swarm. The size may vary, as well as the temporary site. Some land quite high, while others will stay lower where they are easier to hive. New swarms are normally gentle, but even so, the beekeeper should use caution to ensure surrounding observers are not stung. Swarms that have been through a short spell of bad weather can be testy.

# 29

# There are few eggs and larvae in the brood nest

## CAUSE

When a beehive does not thrive, there are many possible causes. Some common reasons are: an underperforming queen, a disease or pest infestation, or pesticide exposure.

## SOLUTION

When the beekeeper discovers a problem in a colony, the first thing to do is to determine the possible causes. The second is to choose a method of treatment or management regime, but a close third consideration is whether or not enough of the season remains for the colony to recover sufficiently, even if the corrective program is successful.

For the beekeeper with several years' experience, a quick review of the combs will usually offer some clues. If any brood is present, is it healthy? Compared to other nearby colonies, is this a feature that is unique to this ailing colony? Is the queen present? If so, is she productive enough to stay on the job? If the queen is not present, a new queen must be introduced. If a queen can be acquired in three to four days, this weak colony's condition can be supplemented during that time by adding a frame or two of larvae and capped brood. Taking time to raise a queen from a larva will take far too much time, so a mated queen will need to be purchased.

Even if brood is transferred and a queen is acquired in just a few days, this colony will need further subsidies of brood and food stores from other colonies. If that is not possible, and the colony is in a dire condition, once you have confirmed that no transferable disease or pest is present, it should be united with another colony.

❋ *The queen in this colony has vanished, leaving the colony without a queen for more than a week. In the interval, the ovaries of selected workers have enlarged enough to produce a few eggs per day. Since the laying worker cannot mate, all the brood she produces will be undersized drones, though they are sexually viable. A colony with laying workers is very difficult to salvage, so most beekeepers combine a laying worker colony with a more populous colony.*

# 30 There is cross comb on the foundation inserts

## CAUSE

**Defective combs built on new foundation are not uncommon, whether it's beeswax foundation or wax-coated plastic foundation. Good combs are never absolutely guaranteed.**

## SOLUTION

Nothing beats a good nectar flow for good comb construction. However, bees can be induced to build comb by supplying them with concentrated sugar syrup. This is a much weaker alternative to a good nectar flow, but the wax produced from the syrup will still be useful for capping cells, queen cell construction, and repairing damaged combs.

There are several occasions when bees build poorly formed combs. Though bees seem to prefer natural comb foundation to wax-coated foundation, poorly built combs on wax foundation are more difficult to reconstruct. The bees sometimes slightly modify wax foundation to build aberrant combs, and those permanent foundational changes are not readily repaired. On the other hand, plastic foundations, sometimes called inserts, are rigid and solid and cannot be modified by comb-building bees. Combs can be scraped away from plastic foundation. After the beekeeper reapplies molten beeswax to the damaged areas, there is a good chance the bees will build normal combs. If this method is used, all traces of the previous defective comb markings on the foundation must be removed or the bees will use those markings as a basis for yet more defective combs.

Sometimes, a significant amount of healthy worker brood will have been constructed on defective combs. Obviously, it would be a waste of resources to destroy this brood. During a warm season, the defective brood combs may be put at the sides of the hive or otherwise away from the brood nest. After the brood emerges, these combs should be either corrected or replaced.

✳ *In this instance, the bees were either on a weak nectar flow or the frame was not properly wax coated. Either way, the comb being produced is unacceptable and will only result in future problems. It should be moved to the outside edge of the hive box. After the brood has emerged all traces of comb should be removed. The frame can then be given back to the bees for them to try again.*

## KEEP IT FRESH

Foundation, either plastic or beeswax sheets, is best received by the bees if it is in new and fresh condition. Older plastic foundation may not be wax coated, while old beeswax foundation occasionally has dents or holes, or is a bit warped. An eager, healthy colony that is in a good nectar flow will most likely build good combs on nearly anything, but fresh foundation is somewhat better.

# The bees keep scouring the hive entrance

## CAUSE

**Referred to as "washboarding" behavior, the reason for this methodical rocking movement is presently unknown.**

## SOLUTION

This is not so much a problem as a mystery. Is this an issue in which the beekeeper should become involved or is this a harmless behavioral activity? It is established in beekeeping literature that bees will apply a coat of propolis and wax around the entrance to their colony or completely around the nest if it is on a rock outcrop or some other exposed site. It is thought that this barrier masks odors and restricts other insect pests from freely entering the honey bee colony. Washboarding bees have been reported to smooth the surface around the entrance and subsequently apply a film of propolis. While this action would appear logical, there is no obvious supply chain for providing propolis to the bees in question. Another assumption is that the smoothing action at the entrance surface serves to eliminate cracks and crevasses where noxious microbes might live, just as polishing the inside of the nest with propolis protects inside surfaces. Certainly, interior surfaces and some exterior surfaces are coated with propolis, but washboarding behavior is not necessarily associated with propolis application.

The behavior does not necessarily occur on the entrances of all colonies, and if it does occur, some colonies commit more effort to the movement than other colonies. Only a few bees may be involved in the activity but occasionally hundreds are involved. In this instance, bees will be roughly aligned in rows across the hive front near the entrance. Since there is no known problem, there is no need for a solution. But this obvious behavior is clearly visible and could be useful information for the beekeeper if the meaning was understood.

## SCOOPING MOTION

Individual bees have been observed washboarding on the inside glass of an observation hive. From this ventral view of the bee, the scooping motion seems to come near the glass surface while the front legs are actively waved in short scratching motions. The front legs do not appear to make contact with the glass surface. Only individual bees have been observed performing this activity, but within the general vicinity, several bees may be performing the behavior.

❋ The bees in this photo are exhibiting behavior known as "washboarding." To date, the reason for this is unknown. It is an intriguing behavior that clearly shows that some common aspects of bee biology are still secrets of the bees.

# 32 Some capped brood cells have small openings

## CAUSE

**Small openings may be temporarily left in capped cells as part of the normal process of covering the entrance, or they may indicate the presence of disease.**

## SOLUTION

If you have noticed punctured brood cappings, you must evaluate surrounding larvae and comb conditions and determine whether they are part of healthy colony behavior or a cause for concern.

As nurse bees cap the prepupa stage of the larva, they begin around the edges of the cell and work toward the center. Stop-and-start efforts may result in a bit of time passing before the task is completed. Near the end of the job a small opening open over the pupa briefly remains before complete closure. The openings tend to be somewhat regular and symmetrical and are generally only visible for a few hours. This is probably the only instance when punctured cappings are a healthy indicator. Alternatively, normal punctures occur in areas where there is a combination of open larvae and recently capped pupae. Unlike unhealthy punctures, healthy punctures are not bunched but will be noticed at random locations—but within the general area of older larvae and capped brood. These are both natural occurrences in the healthy colony.

However, raggedly torn openings in the brood cell cappings could indicate the presence of American foulbrood (see Problem 69), chilled brood or possibly Varroa mite infestation (see Problem 62). Seemingly, these torn openings are expressions of the colony's hygienic behavior that senses the issue and begins the process of uncapping and removing the dead larvae.

## PUTTING A CAP ON IT

Capping both honey and brood is a complex process comprising one of two elements. The first is that several bees work cooperatively to cap a brood cell while the second option is for a single worker to cap a brood cell. Cappings that cover honey are only made of beeswax, while brood cappings seems to be a propolis and wax mixture.

❋ *This is a brood frame of older larvae, most of which have been capped. Small holes remain in cappings that are presently unfinished. In a short time, they will be closed. This is the only instance where these openings are desirable. A ragged, irregular opening usually indicates the presence of disease.*

# 33 Healthy bees are lying dead in front of the hive

## CAUSE

There can be many reasons for the presence of dead bees at the entrance to the hive. Pesticides and diseases are the most common reasons. Other lesser problems related to weather and temperature can also be to blame.

## SOLUTION

Your first step after observing dead bees at the hive entrance should be to determine the dead bees' condition. If the weather was recently cold, dead pupae would mean that developing bees were cold-killed. To help the colony recover, feed both sugar syrup and protein substitute, and periodically monitor its recovery.

If the bees are deformed and predominately without developed wings, this indicates a severe Varroa infestation (see Problem 62). If enough warm season remains, the colony should be treated, but if winter is near, you may need to accept the fatality of the situation.

If the bees are in good shape and large numbers are piled outside, they may have been exposed to insecticide. Check that the queen is not among the dead, and feed the colony to help it recover. If possible, determine where the bees obtained the poison. If the kill was dramatic and all colonies were affected, it may be appropriate to replace the combs.

Dead adult bees in front and bees fighting among themselves is an indicator of robbing behavior. The colony under attack should have all entrances closed with only a very small opening (¾in or so) remaining. Do not reopen this colony for several days and even then just quickly check for the presence of the queen and the condition of the food stores. Feeding this needy colony might restart the robbing behavior. You may need to move the colony to a new location, away from other more powerful colonies.

✳ *A few hundred dead bees out front is not uncommon and indicates healthy hive-cleaning behavior. Particularly during winter months, large numbers out front indicate the bee population is alive and older bees are dying off. Other than assist the wintering colonies with necessary food reserves, there is nothing to be done about normal worker bee replacement.*

## HEAVY LOAD

Typically, Langstroth-styled hives and other similarly designed hives put the entrance near the ground. Heavily loaded departing bees have difficulty gaining altitude and are forced to drop their dead coworkers near the entrance. If the entrance were higher from the ground, the beekeeper would rarely notice the accumulation of dead bees near the entrance.

# 34 The brood combs have turned heavy and dark

## CAUSE

Brood combs naturally become dark due to pigment compounds in propolis, pollen, and cocoons. Additionally, the excrement of new bees seems to darken brood combs. However, it can also occur due to contamination from toxic chemicals in insecticides and miticides.

## SOLUTION

Modern beekeeping has been built on the concept of reusable combs. In the past beekeepers used combs indefinitely. Increasingly, this procedure is being questioned. The recommendation for handling older, dark combs is still evolving, but the following two points are reasonably well established. Firstly, you should mark each new frame with the date it went into use in the colony. This will be useful when frame removal decisions are to be made. Secondly, it is good management practice to replace any old, distorted, dark combs during the spring season with new combs that the bees can rebuild.

The issue of how often frames of comb should be removed is less clear. Some beekeepers do it routinely every five years or so. If pesticide use is heavy in the area, possibly the combs should be removed sooner. If the frames removed from the hive are plastic frames or foundation inserts, the wax can be scraped away and melted beeswax brushed back on. They can then be returned to the bees during a nectar flow period for new comb construction. Wax foundation frames are much more difficult to restore and should probably be discarded.

Most beekeepers tend to replace two to three frames per season rather than ten combs at once. If comb removal is something you want to practice, consider removing all olds combs during the spring season. Then, after two or three years, remove a few frames each year. Start fast and then go slowly.

## COLOR CODED

New combs just a few days old are brilliant white and very soft. After just a few months' use, the honeycombs begin to yellow while areas of combs for rearing brood quickly become a distinct black semicircular shape. Over even more time, brood combs become totally black while honeycombs that never had any brood become more yellow. Both combs, over a year or two, will become much stouter and securely mounted within the frame.

❋ *Comb used by the bees for producing brood typically becomes much darker than honeycombs. In the absence of honey and pollen on this frame, the color change from white to black is clearly visible.*

# There are many dead drones at the hive entrance

## CAUSE

The presence of dead drones at the hive entrance is part of normal colony behavior, but it can also be an indication of food shortage or disease within the colony.

## SOLUTION

In many ways, drones—both alive and dead—are indicators of colony health. Drones are forced from the colony in early fall as part of normal house-cleaning behavior. In this case, there is nothing the beekeeper should do. However, if drones are being evicted during warm months and dead drones or parts of dead drones are on the ground in front of the hive, other problems could be at hand.

If weakened worker bees with deformed wings are mixed in with the dead and dying drones, this indicates a serious mite attack and treatment must be provided or the colony will perish. Indeed, the likelihood of this colony perishing is great. The colony should be monitored for the presence of brood, the queen, and sufficient food stores.

During warm seasons, if both mature and developing drones are eliminated, check the colony for food stores. In times of food shortage, worker bees will eat developing drones and evict mature drones from the colony. This is obviously a food-rationing procedure. A colony with these symptoms should be fed both sugar syrup and a protein substitute.

If white drone body parts are observed at the hive front and if the weather was recently unseasonably cold, this would indicate cold-killed drones. Drone brood is normally placed in fringe locations and is the first to die if cold weather returns. Perform a general hive inspection on the colony to be sure sufficient food is present and to determine any other damage done to the colony. Otherwise, the colony should recover.

# 36

# There are two queens in the brood nest

## CAUSE

During swarming or supersedure episodes, colonies will occasionally tolerate more than one functional queen in the brood nest. This is usually because the colony's queens are in transition and the situation will likely correct itself. The queens are frequently genetically related.

## SOLUTION

While reports of multiple queened colonies are common, there are few reports of this situation persisting indefinitely. Ultimately, one queen will become the colony monarch. When requeening a colony, you should be prepared for this occasional quirk of queen biology. In normal situations, after the reigning queen has been removed, a caged replacement queen is positioned near the brood nest. After four to five days, while the new queen is still caged and she is being introduced to the colony, no eggs should be present. If you return five days later and finds that the bees are flighty, noisy, and aggressive toward the cage, reinspect the colony's brood nest for the presence of eggs. This behavior indicates that another queen exists within the colony. Even if no eggs are present, an unmated queen may exist and the colony would consider itself to have a queen. In this rare occurrence, you should closely monitor the colony's behavior toward the cage.

If no new eggs are present, but the queen's cage is still being treated in a questionable manner, try this: With a small bowl of water at hand, gently release the queen onto a brood comb. If she is immediately attacked and a cluster of angry bees forms around her, drop the cluster in the water bowl and separate the queen from the attackers. Recage her and deal with the extra queen or queens still roaming in the colony.

## CHAPTER FOUR
# MANAGING AND MANIPULATING HIVES

Today's honey bee population lives in two worlds. One is the feral world in which bees find their own nest cavity and make their own way, with varying degrees of success. They lay out their own nest design, provide only natural stores, and swarm as often as the colony can biologically afford to do so. The other world is provided by beekeepers. Artificial domiciles are designed to suit both the bee and the beekeeper. In practice, it is not a perfect arrangement for either party, but the system has allowed honey bees to be kept around the world in a multitude of hive designs.

In this artificial colony nest cavity, some natural conditions are changed to make the site functional. For instance, in many hive designs, the entrance is too near the ground or nests (hives) are positioned in a much greater concentration than they would occur in nature.

The efficient beekeeper must become competent in a wide range of hive management techniques—whether that be helping the colony to prepare for winter, keeping undesirable cross comb under control, splitting colonies that have become overcrowded, or supplying supplements when food is short.

# 37 The bees have built combs in empty spaces

## CAUSE

Given a choice, bees would rather build free-handing combs of their own design. If space is inadvertently provided within the colony, bees will construct natural combs there first.

## SOLUTION

There are several common occasions when empty space is left in the colony: for example, when a frame is removed during queen introduction, when space is made for a feeder within the hive, or when the inner cover is left in the winter position during the spring months. In extreme cases, such as when using early types of plastic foundation that were not wax coated, the bees disliked the foundation so much that narrow, deformed combs were built between the frames rather than construct combs on the foundation. Comb building in empty spaces will only occur during times of a natural nectar flow, or when feeding heavy sugar syrup.

From a hive management perspective, this wastes the colony's resources. The combs must be cut out and the empty space should be filled with frames of foundation or combs.

No doubt due to the energy and resources invested by the colony in new comb construction, the queen will quickly begin laying eggs in new brood combs. By the time the beekeeper discovers the problem, a significant amount of valuable brood may already have been produced. If possible, temporarily tie piece of brood into empty frames and allow the brood to emerge. Overnight, the bees will attach the brood comb pieces to the frames. Once the capped brood has emerged, these frames of comb pieces should be removed and replaced with properly outfitted frames and foundation.

✳ *These natural combs resulted when an internal feeder can was used in late winter, but was not removed before a robust spring nectar flow began. The beekeeper was able to salvage the brood and re-feed the honey back to the colony. The wax was melted into high quality virgin beeswax. However, the bees' energy was wasted.*

# 38 The bees are gathering on the front of the hive

## CAUSE

The honey bee brood nest cannot be allowed to overheat and cause undue injury to developing bees. Using a combination of water evaporation and air movement, bees have extraordinary methods of cooling. When that no longer works, bees will evacuate the brood nest and hang in a mass on the front of the hive.

## SOLUTION

At periods when brood is being produced within the colony brood nest, temperatures must be kept at approximately 94°F. When the temperature rises too far above that level, foragers are redirected by nurse bees to temporarily become water foragers. Along with air movement produced by fanning bees, water is evaporated to cool the brood area and to maintain humidity levels. However, it is not uncommon for the ambient temperature to overrun that cooling mechanism.

At this point it is common for all unneeded bees to leave the nest and hang on the front of the hive. This phenomenon normally occurs during hot summer months when food sources have waned and many worker bees are unemployed. This commonly happens at night too, though this is not as frequently observed by beekeepers. Though this is a natural procedure that occasionally occurs in feral nests, the bees are in jeopardy outside the nest. Rainstorms and the hot sun make life difficult for these exposed bees.

Beekeepers can help by creating other openings within the hive to allow for air movement and also by providing a dependable nearby water source. Also, if the colony is physically crowded, additional hive space can be provided, even if there is no nectar flow in progress.

❋ *The bees pictured on the hive front are mostly older, experienced foragers that have full venom reserves. They are on the hive front because the temperature is too high within the hive. This hive presents a stinging situation. If this hive is to be manipulated, smoke must be used and protective gear should be worn. Additional hive space will allow most of these bees to move inside the hive.*

# 39 The brood nest is overcrowded

## CAUSE

During the winter, bees' natural tendency is to move upward in the hive. The reason for this behavior is not clear. As the season passes, the brood nest will grow at the top of the hive and possibly become restricted in size due to space limitations. At some point, swarming preparations will begin that will be very difficult to reverse.

## SOLUTION

Reversing deep brood chambers is a common way to forestall swarming and to encourage bees to build a strong population. However, this procedure is only necessary if the queen is not laying eggs in the bottom deep in early spring. If she has brood in both chambers, the bees will most likely build in both chambers. Beekeepers can exchange the position of the top deep with the bottom deep to encourage the bees to build the brood nest in the empty space now located directly above the brood nest.

The manipulation should be made during early winter, and the key is to be sure the brood nest is low and honey stores are positioned above the brood nest. Reversing should only be done once, if at all. If the queen is already using both deeps, the beekeeper will have to decide if a brood box reversal will be beneficial. The new space will afford the cluster readily available space to build a larger brood nest, rather than allocating energy to swarming tendencies. Many factors affect swarming, but no doubt lack of space is a major reason that a colony begins swarming behavior. In the spring, give space before it is needed.

❃ *This colony was reversed and proceeded to fill both deeps to capacity. Even with the reversing process having been performed, this colony will still likely swarm. Reversing brood bodies serves to build a larger overall population, but by itself, reversing hive bodies will not prevent swarming. Many other factors must be considered. Swarming genetics and the queen's age are important factors in addition to brood space availability.*

## QUEEN'S MOVE

The condition of the queen is important. A queen that cannot produce the needed brood output will not be made more productive simply by adding more brood space. Alternatively, a queen that has totally filled the upper brood chamber but only has a brood frame or two in the bottom chamber might be more inclined to move into the empty space if it were moved to above the larger hive body.

# 40 The brood nest has been built across multiple boxes

## CAUSE

The colony is small and the nectar flow is weak. Possibly, food stores are also low. Some colonies accept artificial domiciles better than others. Diseases, pesticide exposure, or poor queen production can cause the brood nest not to be established in a compact fashion.

## SOLUTION

The primary issue with this colony that is "chimneying" through the equipment is that it is probably not thriving at this point. There may be a correlation between cluster sizes in relationship to cavity sizes. In most managed hive designs, the cluster is positioned at the bottom of the cavity and grows upward. However, in most feral nests, the colony begins at the cavity top and builds combs out and down from that point.

The beekeeper can help by determining the reason for this behavior. Indeed, the cause could be nothing more than the cluster is a recent split from a larger colony. Compacting the brood nest into a centralized area within the bottom hive body, providing food, and removing some of the extra space are common procedures for encouraging the colony to build a centralized brood nest.

If the queen's performance seems to be the issue, she should be replaced, and if possible, one or two frames of brood can be provided from other healthy colonies. If the incoming food supply is scant, the colony can be fed both carbohydrate and supplemental protein. Beyond those changes, some time will be needed for the colony to show improvement.

## CENTRAL BUILDING

Not only will bees build brood nests in a stilted fashion, but they will, on occasion, put nectar and honey reserves in multiple centralized frames in multiple supers. Commonly, weak flows and below average cluster size may be in play. While no great harm is done, this behavior is a hallmark of a struggling colony. If possible, the beekeeper should address the colony's problems.

❈ *This colony may have just taken the easiest path. Center combs were provided in two deep hive bodies to provide a nest-building site. Rather than produce comb, the colony simply constructed a long narrow nest. In fact, if food conditions improve and the queen is vibrant, this situation may correct itself. However, the beekeeper should monitor this cluster's developmental progress. These issues must be corrected long before winter arrives.*

# 41 The hive is bulging with too many bees

## CAUSE

The colony's hive cavity has become filled with brood, bees, and food stores, and the colony is overcrowded.

## SOLUTION

If a colony becomes crowded, honey crops will be either lost or reduced, pollination activities will decrease, and swarming will probably occur, but the colony will most likely survive. However, a jam-packed colony is difficult to manipulate. The bees have difficulty responding to the smoke stimulus, and bees seem to boil from the colony. With this mat of living bees atop all edges and surfaces, many bees will be crushed during the hive reassembly process. Today, many beehives are kept in residential areas, where bee swarms may or may not be a welcome occurrence on neighboring property. Bees amassed on a hive front will be intimidating to individuals not acquainted with bee behavior.

There are two possible solutions: either provide more space or divide the colony into smaller units. Adding more brood and honey supers will provide more room for the bees to "stretch" out and should help to reduce swarming behavior.

The simplest technique for splitting a populous colony is to make as many splits as there are brood bodies. The hive in the photo was in three deeps, so three single-story colonies would have been formed, of which only one had a queen. After three to four days, queen cells would have started to develop in the queenless thirds. By destroying the queen cells and replacing them with newly purchased queens, a great deal of seasonal time can be saved.

❋ *This is a beautiful colony, but it clearly needs more storage space. Honey production has come to a halt and brood production has been reduced. With little else to do, this colony is very likely to develop swarm cells in preparation for swarming. The beekeeper's investment and energy will fly away. However, the lost swarm will possibly help repopulate the depleted feral bee population, if it can succeed in the wild.*

## SELF-PRUNING

In order to shrink the colony size to fit the cavity size, another option is for the colony to reduce or even eliminate brood production. As the nest cavity fills, increasingly, house bees will store nectar in the brood nest area until the nest become "honey bound." Essentially, the colony prunes itself and the objective of reducing the colony size is met.

# 42

# A large number of bees died off in the winter

## CAUSE
Winter colony kills are usually caused by going into winter with a small cluster, or with inadequate food stores, or both. In mild winters or climates, winter decline is usually related to disease and queen issues.

## SOLUTION
In temperate climates, colony management is always primarily about upcoming winter preparation. The wintering colony does not hibernate. It becomes compacted and quiescent and simply waits out the cold season. A healthy, populous colony can withstand significant amounts of coldness, but if it is sickly or food stores are low, it will be in dire trouble.

The beekeeper can do things to help, but this is best carried out during summer or fall, before the onset of winter. When removing honey for extraction, you should always consider the bees' wintering needs, and leave a secure quantity of honey for the bees. You also need to ensure that the colony is populous and healthy, is headed by a young queen, and that any mite population is suppressed. Sheltering the hive with windbreaks or adding supplemental winter packing can help the colony to maintain optimum temperature. Finally, you can reposition the honey stores so they are above and around the cluster. On warm days during winter months, the hive may be opened briefly while honey frames are repositioned nearer the cluster. Never break apart the wintering cluster: It will not be able to reform before freezing.

Colonies passing the winter in warm climates face different issues. Mild winters allow nearly constant bee flight and possibly some brood production year round. However, there will be few winter food sources, and honey stores and adult populations can quickly decline. Colonies can be provided with emergency feed during winter, but the crisis can possibly be avoided, again, by preparing the colony effectively for winter.

✳ *This colony starved during the winter months. Honey stores were present in the colony but they were not positioned near enough to the hungry cluster. It shows a typical starvation pattern of a tight cluster, with many bees actually in comb cells. The queen is visible near the center of the left margin. Assisting such a desperate colony during the winter, when no honey is present, is difficult and time consuming. It is much easier and much better to make hive arrangements during fall months.*

# The bees are visiting neighbors' water sources

## CAUSE

Bees will seek dependable water sources to supply the colony's water needs. Their search may take them onto neighboring property, particularly during the summer. While the foraging bees are not particularly aggressive, their presence is unnerving to neighbors.

## SOLUTION

If rain is abundant or if large bodies of water are near, water-foraging behavior will not be an issue, but during warm months or in arid areas, bees may make a nuisance of themselves at nearby swimming pools, ponds, birdbaths, or animal drinking troughs. A partial solution is to provide a dependable water source near the apiary and never, ever let it run dry. Even then, some bees will find water sources in places not monitored by the beekeeper.

A common recommendation is to allow a faucet to lightly stream during warm months. However, this procedure is now considered wasteful. Landscape ponds are ideal, but can be expensive. Children's wading pools that are covered with chicken wire and have floating devices have been used with some success.

In general, and for no clear scientific reason, watering devices are located about 50 yards from the apiary. They could be closer. Occasionally, beekeepers use feeders to supply water when other water sources aren't available.

# MINERAL FIX

At times, bees can be seen gathering water from muddy pools or from the drain holes in potted plants. These foragers are seeking trace elements and salts from the soil rather than pure water. In fact, decades ago beekeepers would put a dash of salt in their sugar-syrup mix to encourage the bees to consume the mixture.

✻ *These purpose-made drinkers provide a reliable source of water for bees. The water should be changed regularly to ensure it is fresh, and the equipment kept clean to avoid bacterial growth.*

# 44 The colony looks too weak to survive winter

## CAUSE

**There are potentially many causes for a colony having a small population at winter's onset. Some common reasons are diseases and pests, poor management protocols, or an ineffective queen. In most climates having a true winter, there is little reason to try to winter a small colony.**

## SOLUTION

As winter approaches, the most common solution in cold climates is to combine smaller colonies with larger healthier colonies. There are many ways to combine colonies, but a common and simple technique is to place a sheet of newspaper on top of the healthier colony, perhaps punching a hole or two through the paper. Place the weaker colony on top of the stronger one and give them a few days to eat the paper away. In order for the healthier queen to survive, the queen in the weaker colony should be removed before starting the combining process. If all goes well, next spring splits can be made from the overwintered colony.

Alternatively, if other healthy colonies are available, you could move brood and bees to the weak colony. However, be certain that this will not simply stress two colonies rather than one during the winter season.

In warmer climates, it is possible for a smaller colony to survive the milder winter. You will need to monitor the small colony to be sure that it has ready access to food and that other hindrances, such as skunks or raccoons, are kept away. However, ultimately the warm-weather beekeeper must determine why the colony is small and make necessary corrections.

## WINTER LOSSES

Undersized and otherwise population-challenged colonies are a typical attribute of modern beekeeping. It is a very rare beekeeper who will only have premium colonies all the time. The beekeeper must constantly make decisions about how to help the needy colonies without harming other productive colonies. Winter preparation is paramount in any decision making. It is commonly said that beekeepers should take their winter losses in the fall.

❋ *Deciding to combine a colony is frequently a close call. Maybe the pictured colony can survive if the winter is mild or if the hive is wrapped and winterized.*

# My apiary has been flooded

## CAUSE

To be near an open expanse of nectar and pollen plants, apiaries are sometimes established in remote natural locations where flooding can be a risk.

## SOLUTION

While beehives do not do well while floating in water, it is not particularly unusual for colonies to survive the ordeal. Propolis and wax are both waterproof materials. As long as the hive remains stuck together, there may be areas within the floating hive where the bees can retreat. Everything is based on luck.

If it can be retrieved, flood-damaged equipment can be washed, dried and reused—especially large hive components. Combs will need to be washed free of mud and dirty water before being given back to bees. Pressure washers work well when cleaning hive equipment.

Honey that has been through a flood will not be suitable for human consumption, even if it has not directly been in contact with the water. After cleaning as much as possible, simply giving this damaged honey back to the bees is the simplest plan. The honey will have absorbed moisture and will probably need additional bee processing.

Live bees that have floated down a swollen stream will most likely not be in great shape. Retrieve the bees and give them a few days to settle down. Go through the colonies searching for queens and, as far as possible, clean and clear the brood nest. Once the damage is corrected, the colony will most likely need food assistance. Any seriously damaged colonies may have to be combined with other colonies to accumulate enough bees to form a thriving colony.

# 46

# The hive lower entrance is often blocked in winter

## CAUSE

By natural honey bee standards, the managed hive entrance is generally too near the ground. Not only will leaves and snow block easy access to the wintering colony, as bees die during winter months, the dead body accumulation may be enough to block the lower entrance—especially if it has been reduced to keep mice out.

## SOLUTION

During winter, bees need to exit the colony for several reasons, including to take defecation flights and to die. Bees that have worn themselves out generating heat during winter months will seemingly make a noble effort to leave the colony. Once warm weather arrives, this behavior means less hive cleaning activities for surviving bees.

A general recommendation for colonies wintering in cold climates is to provide an upper entrance for bees to use as an exit during the winter. A common and simple approach for providing an upper opening is to push the top deep back approximately ¼in (0.6cm) so the colony has two upper entrances, one on either end of the colony. This upper entrance also helps ventilate the wintering colony. This probably is not a bad idea for dormant season bees everywhere and not just in cold climates.

In standard Langstroth equipment, the lower entrance is on the bottom board. The upper entrance is either an integral part of the inner cover or it is formed by simply raising the inner cover approximately ¼in (0.6cm). As the wintering colony moves upward during winter months, the upper entrance is predominantly the one used. Different hive designs use different entrance strategies, but in general, all entrances are larger in summer months and reduced during winter months.

# Brace combs have stuck the hive boxes together

## CAUSE

In the natural nest, bees build what appears to be a contorted morass of vertical combs that twist and weave in different directions horizontally. Brace combs (also called ladder combs or burr combs) are common in both managed colonies and feral nests. Over time, these combs accumulate and restrict routine colony inspections.

## SOLUTION

The open space that managed hives must have in order for the beekeeper to remove boxes must be an oddity to comb-building bees. Over time, in response to colony crowding, the spaces between the top bars and the bottom bars in bee boxes (brood bodies, hive bodies, or supers) will be filled in with combs. No doubt it makes for a better hive configuration for the bee, but not beekeepers. These combs are used for drone production, honey and water storage, and as a ladder to move up and down within the colony. Such filler combs are not truly an issue for the hive that is routinely worked, but for hives that are rarely opened, these combs can cause honey to drip and bees to be accidentally killed.

In managed colonies, all that can be done is to respect bee space dimensions strictly (¼–⅜in/0.6–1cm, or just enough space for individual bees to pass each other within the colony) and remove support combs as they are built. Conscientious beekeepers who open their colonies on a regular basis will, by force of habit, scrape away bits of misplaced combs. This routine procedure will keep the pieces small and the disruption to the colony will be minimal. However, many such hive inspections are only perfunctory and the entire hive will not be exposed. Ideally, all of the hive components should be scraped clear about two or maybe three times per year. Two good times to clear these obstructions would be during spring and in preparation for winter.

## A STICKY SITUATION

Bits of combs on the top bars or attached to the bottom bars initially seem inconsequential, but over time, these combs can tenaciously stick bee boxes together. Banging, slamming, and heavy use of the hive tool will be necessary to break the hive apart. These disruptive activities will result in defending bees being alerted and other bees becoming stuck in honey from the broken combs.

✳ *In the hive box shown are filler combs. These small combs are the beginnings of the bees' efforts to tie the top bars to the bottom bars. At this early stage, the combs are easily removed and little damage is done to the colony or the bees. The beekeeper can expect the colony to attempt to rebuild them.*

# 48 The bees have built many swarm cells

## CAUSE

**Swarming is a natural behavior that occurs in bee colonies. The presence of mature swarm cells is a visual sign that a colony has made the decision to split itself.**

## SOLUTION

In the honey bee's natural world, swarming is considered evidence of a successful colony, but in the managed honey bee world, a swarm is viewed as lost energy and productivity. Swarming behavior seems to be initiated by the presence of an aging queen and increased brood nest crowding. The traditional advice for dealing with this is still the best—to requeen regularly and to provide space in the brood nest before the colony needs it. Once swarming preparations are well underway, simply providing brood nest space will not be enough to forestall swarming preparations.

If preparations are already well underway and many queen cells are present, you could attempt to keep the bees in the apiary by making heavy splits from the swarming colony and forcing the bees in the splits to use the swarm cells as queen replacement cells. But even with heavy splits, it is not uncommon for the divided colonies to each cast separate smaller swarms.

Another technique is to destroy every visible queen cell. But missing just a single cell will allow swarming tendencies to proceed. A colony with a strong proclivity to swarm will actually swarm without a swarm cell in the parent colony. In such cases, the parent colony is most likely doomed to a short, queenless life.

* *Swarming is considered an indication of a successful colony in the natural world, but to beekeepers it represents a loss of energy and productivity.*

## SWARM BEHAVIOR

A swarm that has only recently landed (pitched) is normally a gentle bunch of bees. At that moment, they are homeless and will frequently accept any dark domicile that is offered to them. A proficient beekeeper should know how to predict and manage a swarm effectively.

# 49 Rain has stopped pollen and nectar flow

## CAUSE

**Rain showers, even light ones, temporarily eliminate the pollen and nectar flow for foraging honey bees. Both nectar and pollen are rinsed away. Even if it were not, flying in rainy weather is difficult for bees.**

## SOLUTION

While some rain is clearly necessary, rain at the wrong time or too much rain at any time causes problems for agricultural production, including beekeeping. During some spring seasons, there is simply too much rain or poor weather for bees and plants to work together. No beekeeper can control the weather, but in some instances, there are ways that you can assist rain-restricted colonies.

Recent colony splits or any colony that is light in food stores will readily use supplemental food if supplied by the beekeeper. This assistance sounds simple, but if the rain is steady, opening colonies to provide food can be somewhat traumatizing to the colony. Ideally, you will have already installed some type of feeder. A hive top feeder is the most convenient. While some pollen supplements can be fed into the top feeder, this product is usually best consumed by bees if it is put just above the brood nest. This will be a wet, unpleasant task, but once completed, the needy, confined colony will appreciate it.

When the rainy weather does break and a nice day ensues, you will probably be eager in open up the colony to assess its condition—especially if you have also been confined indoors by the weather. But if the colony does not need assistance, it is better to let the colony forage on that clear day and recover from the previous confinement than to interrupt work with an inspection.

## GO WITH THE FLOW

Novice beekeepers often ask how to tell when a flow is on
and in what ways preparation should be made for it. Local
beekeepers will know the average dates that primary flows
begin and end. Watch the progression of the primary crop
blossom for a clue. But when the nectar flow begins in
earnest, it will be obvious. New honeycombs will appear nearly
overnight, and thin, watery nectar will be in any available comb.

✳ *Bees struggle to fly in
rainy weather. Even a
light shower is enough
to temporarily rinse
away pollen and nectar
from flowers.*

## CHAPTER FIVE
# QUEEN PRODUCTION AND MAINTENANCE

All aspects of bee biology are critical, but the queen's functions within the colony are particularly important in several ways. She serves as the egg producer and in warm months can produce up to 2,000 eggs per day and maybe a million eggs in her lifetime. In conjunction with the drones with which she has mated, the queen serves as the genetic reservoir for the colony. Various biological characteristics such as colony defensiveness, worker color, and hygienic behavior are determined by the genetic traits derived from the queen bee and the drone semen she has stored within her body.

An important part of queen management is assessing if and when a queen needs to be replaced. Some of the reasons why this might be necessary are covered in this chapter, along with related issues that might arise, such as temporary queen storage, and how to introduce a new queen to the colony successfully. Though she may be the biggest bee among her peers, that doesn't mean the queen bee is always easy to find; tips for keeping track of the queen are given, as well as how to cage her once she has been found. Queen production is a broad topic in itself, and problems related to queen grafting and enforced in-colony production are also looked at here.

# 50 I need to store a replacement queen

## CAUSE

**Sometimes queens arrive from the producer when the beekeeper is not able to use them immediately. The weather is an important factor as to when a new queen can be introduced, or there might be a delay while the beekeeper searches for the old queen inside the hive.**

## SOLUTION

Mated honey bee queens can be stored from just a few days to as long as six months, though they should be released into colonies as soon as possible. There are many ways to store queens. A common method is to provide caged queens with a food source and cover the queens in a blanket of young nurse bees that will feed and warm the confined queens. The process occurs in a special box that is partially screened for ventilation. More young bees are added as needed and a water-soaked blotter is used to provide water to the bees and queens.

A single queen can be stored for a few days in the shipping cage in which she was mailed. She should be kept calm in a cool, dark location. If she has attendant bees confined with her, it would be helpful to remove any that have died. The queen should be given a single drop of water about once every day or two. If you realize that more time will be needed, the caged queen can be stored for a few more days near the brood nest in an active colony.

Queens are stored long-term (banked) in modified colonies. The primary storage colony requirements are that there is no free-ranging queen within the colony; that no worker attendants are within the stored queen cages; and about once week, the banking colony must be given brood and bees from other colonies to keep abundant nurse bees present (either as capped brood with no young larvae or newly emerged worker bees).

❊ *These queens in white plastic cages are being temporarily stored in a ventilated box. The orange bar supports and spaces the individual queen cages. The storage box has both water and sugar syrup available to the nurse bees. About every other day, new nurse bees are added that will accept the caged queens and feed them.*

# 51

# I don't know where to look for the queen

## CAUSE

**The queen moves around the colony and could be absolutely anywhere—even on the inner cover.**

## SOLUTION

A methodical approach is required when searching for the queen in a hive. As beekeepers acquire experience, they develop a skill for seeing the queen in the midst of many other worker bees, but this is not a skill that works every time, and if the queen needs to be quickly located, it can be exasperating.

Open the colony quietly and search for brood frames with eggs and very young open brood. Scan the frame slowly and consistently on one side, then gently flip the frame to scan the other side. In the slot that results from one frame being removed, quickly check the exposed sides of the combs that are still in the colony. The queen will move away from the light and activity so she may just be staying one frame ahead.

When searching for the queen, it is helpful to break the hive into it component parts and search for the queen in each box. Don't forget the hive box walls. If the hive is not disassembled, the queen will move from box to box and become even more difficult to find. Another method is to place queen excluders between hive components and wait three days before examining. The box containing eggs will obviously be the box housing the trapped queen. Finding the queen on demand is a challenge for beekeepers at all skill levels.

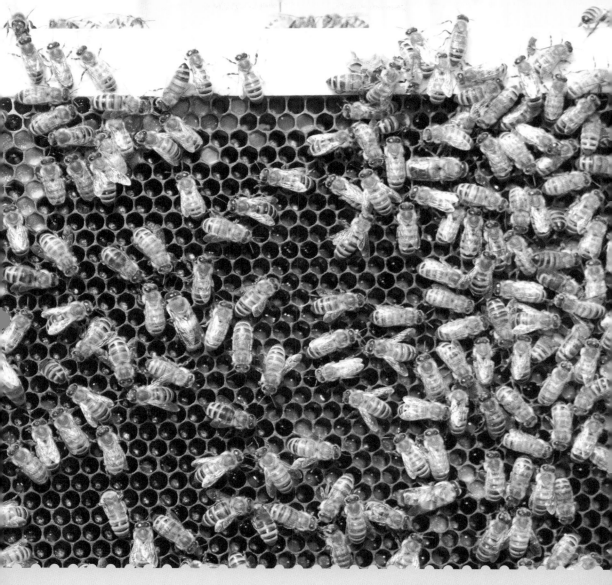

## LOST IN THE CROWD

While a hive box is being examined, occasionally a noticeable mat of bees will develop on the top bars of a box that is not being examined at the time. This could be a clue that the queen is in that area and the disturbed bees are rallying to her location. The queen is not always there, but it's a good place to start when examining that box.

❋ *The queen seems to be able to hide in plain sight. While the eye is drawn to the bulk of the bees on the comb at the right side of the frame, the queen in the photo is at the far left.*

# 52 The queen is not a good egg producer

## CAUSE

Other than advanced age, there are several issues that can restrict the queen's biological performance, such as low drone populations and poor weather for mating flight.

## SOLUTION

If you bought your new queen from a reputable producer, the colony may be out of balance for other reasons. If possible, provide a frame or two of capped brood from a healthier colony and feed carbohydrate and protein to be sure that food is not the issue. Secondly, ensure that Varroa mites or other bee maladies are not the cause. Bad weather or pesticides may also be suppressing the colony's growth. If indications strongly suggest the queen is to blame, replacement is the norm—provided it's not too late in the season to be of benefit.

When replacing a queen, you have two options: to buy a replacement or produce your own. During late spring through late fall, queens can be shipped by mail or purchased at bee supply companies. Queens come without a guarantee so you should buy from dependable producers (the new beekeeper should consult more experienced beekeepers).

Colonies can produce their own queens, but it takes around fifty days, by which time much of the warm season will have passed. You can speed things along by modifying a colony to produce queens in four steps:

- De-queen the colony.
- Remove all eggs and very young larvae.
- Destroy any natural queen cells that have been started.
- Transfer a very young larva to an improvised queen cup and give this improvised queen cell to the queenless colony.

The first three requirements are simple, but transferring larvae (step 4) requires practice (see Problem 56).

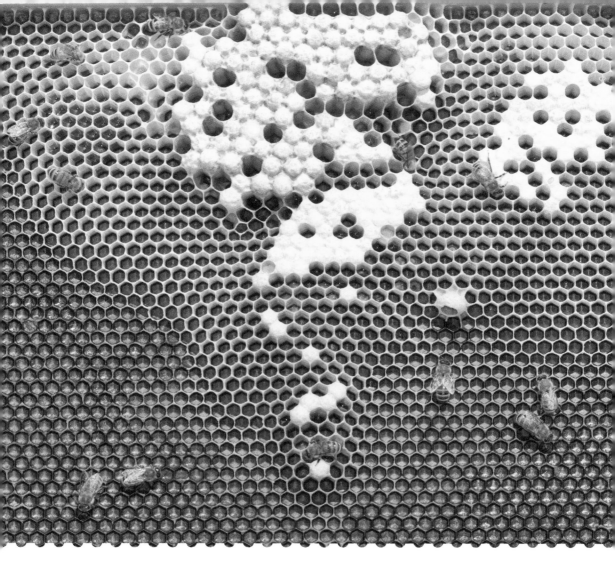

## SLOW INTRODUCTION

Since replacement queens are in no way related to the colony, they need to be introduced to the population slowly. Normally a sugar plug confines the queen in her cage. Worker bees slowly eat the plug, clearing the opening for the queen to escape into the colony. During this time, younger bees will have been feeding her, allowing her to acquire the colony's odors, thus making her more acceptable.

❋ Is this a poor queen or is this a colony that is having other bad luck? The presence of a queen cell is concerning. But the queen has produced both worker and drone brood. Combs are being slowly produced, indicating a lackluster nectar flow is underway. The beekeeper must decide if the problem lies with the queen or not.

# 53 Some cells have multiple eggs

## CAUSE

When multiple eggs or larvae are spotted in one cell, the worst possibility is that the colony has become queenless, and laying workers are emerging. A lesser event is that the queen has deposited two or more eggs, resulting in the cell having too many developing larvae.

## SOLUTION

If the queen mistakenly lays multiple eggs in the same cell, for a short while there will be multiple larvae in the cell. Nurse bees should not allow this to go uncorrected, for there is only room for one mature pupa in a single cell. The extra larvae will most likely by removed and eaten by nurse bees. Little involvement is required by the beekeeper, who may even have no idea that the issue even exists.

Things are much more serious if the problem occurs because the colony has become queenless and has no way to produce a new queen. Depending on the species of bees, workers that are no longer being suppressed by an active queen's pheromones will develop enough egg-laying ability to produce a few unfertilized eggs. These worker-laid eggs will all develop into drones. While a drone population is necessary in a healthy colony, excessive numbers of drones are not beneficial. There are too few workers to carry out the colony's tasks, and the colony withers. Such a depleted colony is difficult to requeen because the colony seems to consider itself queenright and will not readily accept a new one.

While a laying worker colony can be saved, its future is not good. The colony is severely weakened, and in most areas, there will not be enough warm season remaining for it to recover its health. Traditionally, a sheet of newspaper is placed on top of a neighboring colony, and the laying worker colony is placed above the newspaper. The two colonies combine.

## HOMEWARD BOUND

A common solution for a laying worker colony recommended in the old bee literature is to move the afflicted colony a few hundred yards away and shake the bees out. It was thought that the laying workers—being like queens—could not find their way home. This is untrue. Laying workers will readily fly back to the original location.

✳ *When pictured, the typical photo of laying workers shows multiple eggs within single cells. This is an easy diagnosis. More rare is the photo of a single cell containing multiple larvae shown in the photo. In the lower left corner bottom row of cells is a cell containing three larvae. This over-occupied cell will need some nurse bee attention. There is only room for a single adult bee.*

# 54 Finding the queen takes a long time

## CAUSE

**Finding a single insect among thousands of other similar insects is not easy. Additionally, the queen will actually hide or move to the other side of the comb. Since they look much like workers, finding a small virgin queen is even more difficult than finding a mated queen.**

## SOLUTION

Queen marking is a technique that has long been used by beekeepers to make it easier for them to pick her out of the crowd. When queens lived longer, beekeepers would clip the tip of a forewing just a bit. It was thought that a clipped queen would not fly far or high when leaving with a swarm. Also, so her age could be told, individual beekeepers worked out a personal system for clipping a left wing on odd years and the right wing on even years. Since the process was thought to encourage supersedure, that form of marking is no longer used.

Small spots of enamel paint are now the preferred way to mark a queen. At hobby stores, paint pencils are available with a paint-loaded tip that greatly eases the marking process. Small bottles of model-maker's enamel paint, as well as fingernail polish, can be used as queen-marking paints. The problem with all of these above is that quite a few minutes are required for the paint to dry. Only a tiny mark should be applied directly on the queen's thorax and not on the top of her head or smeared on her wings and abdomen. Marking drones is an easy way to practice this procedure. Lightly applying the mark in a circular fashion, rather than applying a dollop of paint, provides a thinner coat that dries quicker.

Happily, queens rarely sting when being held for marking so that threat should be reduced.

❋ *In the United States, the queen color code is: Year ending with 1 or 6—white, 2 or 7—yellow, 3 or 8—red, 4 or 9—green, and 0 or 5—blue. Beekeepers can use this system for determining when a queen was replaced or needs to be replaced.*

## MAKING YOUR MARK

A brush is not a good tool to use to mark a queen bee. A small dowel rod or broken toothpick dipped in the paint that remains in the lid of the paint jar provides just about enough to make an obvious but small mark on the queen's thorax. Colored adhesive numbered plastic dots have also been used for many years as a queen-marking technique.

# 55 The queen is difficult to cage

## CAUSE

The entrance holes in most queen cages are barely ⅜in (1cm) in diameter. The queen is not eager to be put in a cage, and it is important not to harm her in the process. If the queen is dropped, she may take flight.

## SOLUTION

Putting a queen in a queen cage does require practice and confidence, but a few suggestions may help. Make sure propolis and sticky wax has been removed from your hands before picking up a queen. It is sometimes helpful to work in front of a bright but closed window. If the queen does escape, she will head toward the glass, where you can retrieve her. Indeed, when marking the queen, some beekeepers let her fly on the glass until her paint is dry. Due to all the cracks and crevices, a technique that is a bit more risky is to attempt to cage the queen while sitting in a vehicle with the windows and doors closed. If the queen escapes, she will fly to the windows. Be sure all ventilation holes and vents are closed or the queen can get lost in the automobile's air-handling system.

The queen's thorax is surprisingly rigid. If possible, grasp her by both of her front wings and transfer her to your opposing hand and hold her with your thumb and forefinger. If you make an effort to grab her and miss, she will surely become skittish and begin to run about and use her wings as if preparing to fly. Stay calm and try again. Once you have her, she will be squirmy and do her best to escape. Make sure that your grip does not become too tight. Always focus on the thorax and not the head or abdomen.

## A USEFUL SKILL

Queens may need to be caged for a variety of reasons: you may have to cage a queen found in a swarm in order to prevent the hived swarm from leaving again, or perhaps you wish to contain a queen in order to sell or give her to another beekeeper. If a queen is released too soon and attacked by house bees, you will need to recage her. All in all, caging a queen is a useful skill for beekeepers to perfect.

✻ *A queen bee can be kept in a cage until the time comes to release her into the hive.*

# 56 Larvae won't slide off the grafting tool

## CAUSE

**The larvae required for queen production are hardly the size of a honey bee egg. The process also requires that each tiny larva be moved to a different cell without injury and positioned on the side it was originally on. Practice and manual dexterity will be required.**

## SOLUTION

To help with this task, use a grafting tool with a thin, flexible tip. Such tools can be purchased from supply companies or can be simply made scraping a toothpick-sized piece of wood to a very thin-edged tip. Other materials commonly used are goose quills or very thin plastic strips.

Rather than attempting to move the entire larva onto the tool, approach the back of the C-shaped larva and leave both ends hanging from the edge of the grafting tool. When moved to the queen cup, these "dangling" ends will snag the cup bottom and pull the small larvae from the grafting tool. If the larva is completely on the tool edge, it will have to be pushed from the tool, which will most likely damage the larva. There are also devices available that do not require you to handle the small larva. But all devices used to produce queens require some finesse and attention to detail.

A surprising quantity of healthy bees will be required to produce and feed the huge amounts of royal jelly that the bars of queen larvae will require. When producing queens, most colony resources are directed to young bees and specialty hives for growing cells and mating the emerged queens.

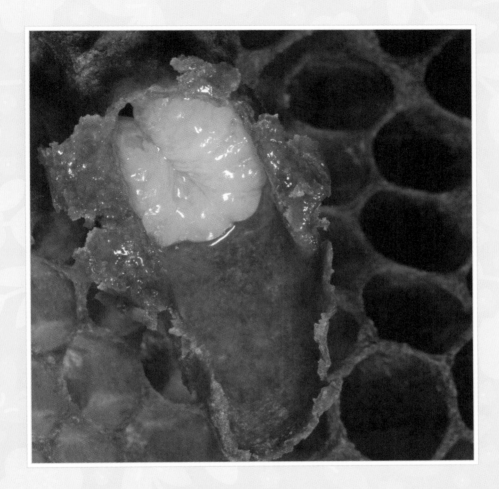

## THE RIGHT TOOL FOR THE JOB

Queen larvae are not grafted in the truest sense of the word. They are transferred to wax cups or plastic queen cups. The trick is in how the larvae are moved from one cell to another. Small, flat pliable tools have been improvised from toothpicks, brazing rods, twigs, and goose quills. Manufactured grafting needles are available, but there is no standard grafting tool.

❋ *This plump queen larva began life as a tiny egg. Trained beekeepers can transfer a larva no bigger than the original egg to specialized queen cups. Nurse bees are then encouraged to feed the queen larva for about five days as it undergoes spectacular growth.*

# 57 Too few drones are available for queen mating

## CAUSE

Due to its longer development time, the Varroa mite preferentially attacks drone brood. Even if a drone survives, the blood feast that Varroa mites take weakens it. Additionally, there are few beekeepers who intentionally provide drone brood space within the colony.

## SOLUTION

Each successfully mated queen requires about 7–15 mature drones. In the past, beekeepers could depend on wild colonies to produce drones, but these colonies are now rare. You can supplement drone populations by providing drone brood foundation and allowing healthy colonies to produce these male bees. For identification purposes, commercial drone cell foundation is green. If the frame is a plastic frame/foundation combination, it too is all green. If green drone foundation is installed in wooden frames, green paint marks should be made across the top bars to indicate that it is a drone frame. Additionally, beekeepers can leave drone brood on frames or combs rather than systematically scraping it away.

Unfortunately, a conflict is at hand. While these green frames produce significant numbers of needed drones, if other Varroa controls are not used, the mite population will grow within that colony. In fact, a common method of suppressing Varroa populations is to use drone brood as a trap crop. Just as the drones are being capped, remove the frames and freeze them, thereby killing both drones and parasitic mites. The frame can then be returned to the colony, where bees will clear the frame of dead mites and drones and once again begin to raise drones.

Modern beekeepers need to encourage drones without increasing predacious mite populations. Monitor the colony's mite population and employ control measures when needed. Otherwise, drones should be allowed to develop to maturity to assist with queen-mating procedures.

## DRONES IN FLIGHT

Inside the hive, drones appear lethargic and sluggish, but outside they are artful fliers—fast and assertive. Drones make their way to flight areas called drone congregation areas (DCAs), where they await the rare arrival of an unmated queen. It has been estimated that only 1 percent of all drones ever successfully mate with a queen. After maturity, drones are not specifically assigned to a single colony but may drift from colony to colony.

✳ *Drones hatching from brood cells. Each successfully mated queen requires 7–15 mature drones.*

# 58 Some of my queen mating nucs have failed

## CAUSE

Nucleus colonies are small colonies that have limited resources and are unable to rebound if hit with various hive maladies such as diseases and pesticides. All in all, small splits or nucs will need a lot of attention.

## SOLUTION

For practical purposes, queens are mated in small—sometimes very small—colonies. In this way, many more queens can be mated using the same quantity of nurse bees. For instance, rather than using one large colony to produce a single queen, many smaller units, called mating nucleus colonies (nucs), can be made from the resources of the larger colony. Each of these small units requires a mature queen cell. If left to their own ways, few, if any, of these colonies can build enough resources to survive the next winter. Indeed, they are never intended to survive the winter. As the hot summer approaches, and robbing increasingly becomes an issue, these small units should be recombined into a larger colony.

In order to provide consistent food to the mating nuc, you will need to constantly feed syrup and pollen substitute. If it does not thrive, add bees and brood from other more successful colonies. In hot climates, mating nuclei (also called baby nucs) should be placed in the shade, while in cooler climates they should be in full sunlight. Vermin such as raccoons and skunks can be very hard on these small colonies, so place them on stands or provide protective enclosures.

✳ *A fully developed queen cell that has been produced naturally within a hive.*

## MATING NUC SUCCESS

While some mating nucs fail, many do not. From them come fully mated prolific queens. Because they essentially have only a small defensive force, bees from these small colonies do not sting much. The queens are easily found because there are only a cup or two of bees in the colony. They only need to be successful for a couple of cycles to have paid for themselves.

# 59 My replacement queen was killed

## CAUSE

**Even if they have no hope of acquiring a queen on their own, older honey bees will readily kill any replacement queen that they suspect as a colony invader. If a new queen is released too soon, hostile bees will likely kill her. Foreign unmated queens are particularly unacceptable to older workers.**

## SOLUTION

To avoid losing a replacement queen to hostile bees, her introduction to the colony should be gradual. The colony ideally should be allowed to stay queenless for a day or so before a new queen is offered to the colony. If practical, you should remove any attendant workers that are in the cage, and place the newly caged queen as near to the brood nest as possible. The recommendation used to be that the queen could be released in a couple of days, but this advice is now largely disregarded. Generally, the queen should remain caged for about five days.

When the caged queen is presented to the bees, they will show an immediate interest in her. The bees will treat the cage as they would the queen if they could just get to her. If you see the bees massed and tightly clinging to the cage, the queen inside has not been accepted. If the bees use their scent glands and are easily brushed away from the cage, she can be released.

When working with queens and handling them, be calm and gentle. If a newly released queen becomes excited and runs, she will draw even more attention to her presence.

❋ *If a caged queen is released before the colony has accepted her, hostile bees will likely kill her. Recage the queen immediately at any sign of aggression.*

# CHAPTER SIX
# DISEASES AND PESTS OF HONEY BEES

As is true with every other living organism, honey bee colonies must deal with diseases and pests. Colonies that are generally healthy can withstand the occasional onslaught of disease pathogens, but they may become overloaded, and when this happens it is the beekeeper's responsibility to take decisive action.

Some diseases and pests are worse than others. Varroa mites are a challenge to bee colonies in much of the world. Only recently, scientists have begun to explore the effects of the viral load that is transferred to honey bees by Varroa as they feed on the bees' blood. The more established diseases, such as American foulbrood, are better adapted to bees and do not normally cause the devastation that an unrestrained Varroa mite outbreak can. The key to effective disease and pest control is in recognizing the signs early on and knowing how to deal with them. As the problems discussed in this chapter show, it is not always clear cut; the symptoms of some diseases are similar, and a colony may be suffering from more than one malady at once. With knowledge and experience, though, you can grow adept at reading the signs.

# 60 There are mice living in the beehive in winter

## CAUSE

During fall months, mice begin to search for a suitable site for overwintering. Beehives—either empty or occupied—offer a protected wintering location. Dead bees, residual honey stores, and unused pollen are nutritious food sources for adult mice and their growing families.

## SOLUTION

There are various reasons why the presence of mice in a beehive is undesirable. Mice agitate the bees, causing them to consume larger amounts of winter honey stores. When building nests, mice cut the combs and chew frame components in order to make a cavity. Additionally, the buildup of mice droppings creates an unwholesome odor. All of the above makes a very strong case for blocking any entrances that mice could use to gain entry to the hive.

In one way or another, all entrance reducers close down the hive entrance to ⅜in, which mice cannot get through. Some models also reduce the entrance width. The simplest procedure is to modify the bottom board so the bottom entrance is set to ⅜in (1cm) all the way across the hive front. Alternatively, a ¾in (2cm) hive entrance can be closed down with a wood strip that fits tightly in the hive entrance and has a notch cut in the middle that measures ⅜in × 3in (1 × 7.6cm). Many beekeepers use a 4in strip of ¼in hardware cloth, folded to 90°, that fits tightly on the hive front and the bottom board and completely covers the entrance. To prevent animals from removing it, this strip should be lightly stapled in place. While live bees can readily pass through this wire grid, house-cleaning bees cannot remove dead bees from the hive until it is removed in the spring. Commercially manufactured entrance reducers are available that are adjustable so they will fit any hive front and are easily installed.

## MOUSE PROOF

In storage, hive equipment should be stacked on a solid surface such as a cement floor. Alternatively, outer covers can be inverted and used as a tray in which equipment is stacked. On top of the stack, another outer cover should be used to prevent mice from entering there. Make sure there are no cracks or rotted corners in the equipment that could allow mice to enter. Such holes must be closed with metal screening.

❋ *These hives in Slovakia have the larger summer entrance reduced to ⅜in to prevent mice from entering the dormant hive during winter months. Another more modest benefit is that cold winter winds will be restricted.*

# 61 There are multiple problems with the colony

## CAUSE

During a difficult period, the colony can succumb to more than one malady. A bit of diagnostic expertise may be required of the beekeeper.

## SOLUTION

Expertise in honey bee disease and pest identification can only come over time. While photographs and other educational materials are readily available, it is likely that you won't experience a disease firsthand until it presents itself in your colony. If there are multiple problems, it can be even trickier to determine how to treat the ailing colony.

Competent disease recognition is absolutely necessary to maintain vibrant, productive colonies. Stay abreast of disease reports, discussions at meetings, and updates published in the beekeeping literature. Importantly, you should develop a cadre of beekeeper friends with whom you can discuss problems or ask for opinions. It is in their interest to help you, as bee diseases can spread to other hives in the local area if not dealt with effectively.

If you are new to the craft, regulatory or university staff may be able to help with a complex, mixed diagnosis. If the colony has time and energy, it may once again regain control of its own health. Bear in mind that while you are awaiting confirmation of the diagnosis, some of the earlier symptoms may fade.

Beekeepers should always be looking for problems—anywhere, even outside the colony. Inside the hive, watch for low-level problems and address them before they become major problems.

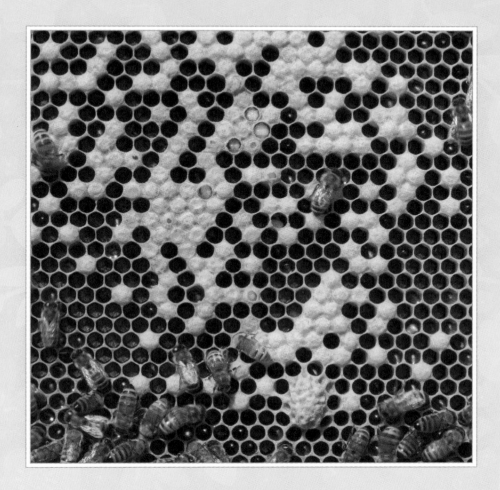

❋ There are multiple problems with the pictured colony. There is a single Varroa mite in the upper left quadrant of the photo. The developed pupae (top, center) should not be uncapped at this time. Near the bottom center is a dead prepupa. There is a poorly formed queen cell near the bottom center of the photo and there are no uncapped brood or pollen stores for future uncapped brood. This colony could use some help.

# 62 Varroa mites have overrun the colony

## CAUSE

Varroa (*Varroa destructor*) is a voracious external pest on honey bees in some areas. If left untreated, they can quickly build to population levels that can kill a populous colony.

## SOLUTION

There is no chemical or management procedure that will completely eradicate this pest, so individual treatment regimes must be developed. One method is drone brood trapping. Drones require approximately 23 days to mature, while workers require just shy of 21 days. Apparently due to the longer development time, Varroa mites preferentially seek out developing drones. You can therefore use drone combs to attract mites away from other areas of the brood nest. Once the comb is filled and the drone brood is mostly capped, remove it and freeze it. Both drones and mites will be killed. The comb can be used again for the same purpose. During the warm season, you should perform this eradication procedure about every 18–20 days.

Manufactured chemicals are available to suppress Varroa populations. Most are applied on plastic strips that are hung between the frames, or as blotters that are laid on top of brood nest frames. Times and length of application will depend on the product, so follow the instructions carefully; always wear heavy plastic gloves and avoid inhaling the vapors. Chemical control materials should be rotated annually in order to minimize development of resistance, and integrated pest management should be practiced: mite resistant queens, drone trapping, and screen bottom boards should be used alongside.

Unless the product label explicitly allows it, do not use chemicals when honey supers are on the hive. Consider periodically replacing brood frames to avoid chemical residue accumulation.

❋ *The colony pictured has died from a massive Varroa mite infestation. The red tortoise-shaped mites are Varroa mites. A bee with deformed wings is present, indicating the presence of a pathogenic virus. This attack occurred during early September. The afflicted colony had no time to recover.*

# 63 There are ants nesting beneath the outer cover

## CAUSE

Depending on the ant species, some kinds of ant may actually nest within the colony. Other species live outside but near the colony. In general, beehive environs are good areas for ant colonies to set up housekeeping. There is a steady source of food from the hive and in some cases protection from weather.

## SOLUTION

Ants can cause damage within the hive. Species that tunnel in wood, such as carpenter ants (*Camponotus spp.*), can damage equipment in a manner similar to that of wax moths; they are also reported to damage high-density polystyrene beehives.

As ants and bees are related insects, in general, any chemical insecticide that kills or repels ants will probably have the same effect on honey bees. Be very careful when using commercial ant insecticide near bees.

Cinnamon sticks and cinnamon powder are traditionally used by beekeepers as a repellent. No doubt this common food product will work at times, but not at others. Boric acid mixed with sugar or oil is frequently reported to work. The acid acts as an abrasive on the cuticular surface of the bee's exoskeleton. Essentially, the insect loses its ability to retain moisture. However, if they are not protected from the concoction, bees will experience the same harmful effect.

In years past, beekeepers sat the legs of their hive stands in cans of oil or water, in the hope that this would serve as a moat barrier. Sometimes it did, but at other times the ants devised ways to bridge over the moat with leaves or twigs.

These are just some of the proposed methods for controlling ants in the hive, but none is universally accepted as the "best" way.

# 64 Small hive beetles have invaded the colony

## CAUSE

The newest hive pest to establish itself in the United States (1998) is the small hive beetle (SHB, *Aethina tumida Murray*). Adults lay eggs in bee colonies. The larvae develop in large numbers, foul honey stores and combs, and stress bees. Bees frequently abscond, leaving behind combs and equipment that are difficult for the beekeeper to clean.

## SOLUTION

This pest does not universally affect all beekeepers in all areas. In its home range in sub-Saharan Africa, it is not considered a major pest of honey bees. Research efforts have not yet produced an effective chemical control.

Various traps have been developed that do successfully harass the SHB, though they won't eradicate the problem completely. These traps work in two ways: they form a tight space, which the beetles prefer because they feel safe from bee attacks in confined areas; they are filled with vinegar, which acts as an attractant and may also drown some of the adult beetles. Two traps are usually placed between brood frames near the center of the nest for each brood box. The lightweight plastic traps fit flush between frames and are easily installed, and can be discarded when filled with dead and dying beetles.

The only chemical controls are GuardStar™, which is applied as a ground drench in combination with Checkmite™ strips placed beneath pieces of corrugated board. Beetles hide beneath the board and come in contact with the strip, which has coumaphos as the active ingredient.

Beekeepers have reported that beehives in the sunlight survive better, but managing bees in the hot summer sun is uncomfortable. The best control advice is to keep the hive healthy and strong, keep the bottom board clear, and don't disturb the bees more than necessary.

# There are signs of AFB in the hive, but no foul odor

## CAUSE

Bee parasitic mite syndrome (BPMS), a Varroa/virus interaction, looks very similar to American foulbrood (AFB), but without the foul odor and oily-looking cappings. It is caused when a large population of Varroa mites overruns a large bee colony.

## SOLUTION

A colony experiencing bee parasitic mite syndrome (BPMS), a Varroa/virus interaction, is most likely doomed. Virus particles that have always been associated with bees have found a much improved transmission route with Varroa. Varroa does not always kill bees and brood, but it does weaken them. Varroa-assisted virus transmission has upset the evolutionary balance between the viruses and its honey bee host. The virus infection causes more damage than Varroa feeding damage.

The only solution is to prevent the mite population from reaching such high levels within the colony in the first place. Keeping mite populations reduced within the hive and performing these control procedures in timely and systematic ways is the best way to prevent BPMS. There are multiple control procedures that will suppress Varroa populations, but none will eradicate the pest.

Common methods for sampling the current population of Varroa mites are: sticky bottom board screens, ether rolls, or sugar shakes. Each method will yield an estimation of the number of Varroa contained in the sample. It is difficult to present the standardized threshold number at which point the beekeeper should initiate treatments. Speak to fellow beekeepers about the local threshold number.

✽ *BPMS strikingly resembles American foulbrood. Some adult symptoms are: Varroa mites are obviously present, crawling bees, littered entrance board, and a spotty brood pattern. However, the dead larvae do not rope out, scale is not brittle, and it has little odor. American foulbrood has an obvious odor and the cappings have an oily appearance. Additionally, scale will be brittle and firmly attached. Importantly, the comb can be reused, but American foulbrood combs cannot.*

# 66 Chalkbrood is affecting colony productivity

## CAUSE

Chalkbrood is caused by the fungus *Ascosphaera apis*. The disease is found essentially worldwide.

## SOLUTION

Severe Chalkbrood infection results in the dramatic decline in productivity within the afflicted colony and reduces the adult bee population. Even if the colony recovers, it will not thrive during the remainder of that season. The time-honored solution to Chalkbrood is to requeen the colony, thus giving the colony enough of a brood break to remove the diseased contaminants from the colony. It is also thought that the new queen's stock will provide more genetic resistance. It remains unknown how often a colony is actually improved by this procedure, but there is little else the concerned beekeeper can do. A few beekeepers have reported dramatic declines in the occurrence of Chalkbrood in irradiated equipment; however, most do not have access to such a facility.

Chalkbrood spores are remarkably persistent and may remain viable for 15 years. The spores can be spread from hive to hive by bees that are either drifting or robbing in combs and equipment that the beekeeper provides to the bees. These spores seem to be present nearly all the time and are not expressed until proper conditions are met. Moisture is a commonly accepted condition that encourages the appearance of the disease, but it is readily apparent in hot, dry climates, too. It is therefore vital that beekeepers do what they can to guard against the spread of spores. Wipe the hive tool and smoker with a disinfectant to prevent transferring spores to uninfected colonies. If possible, don't use your gloves when inspecting an infected colony and clean your hands afterwards. Monitor frame exchanges so that nearby colonies are not exposed to fresh Chalkbrood spores.

## MUMMIFIED

As the name suggests, only the brood is affected by Chalkbrood. Nurse bees distribute the spores in brood food, and the affected larvae die in the capped cells. When removed by the nurse bees they initially appear bright white and fluffy. As the dead brood dries it forms a hardened mummy that grows darker as the spore-producing phase begins. If the disease is rampant and the comb is shaken, mummies remaining in the combs can be made to rattle.

✳ *Mummies quickly pile up around the hive front. Inside, on the bottom board, large numbers of hard, darkened mummies will also accumulate. Each of these hardened pellets is the remnants of a honey bee that died during development.*

# 67 Wasps are attacking the bees

## CAUSE

**Dead or dying bees are a tempting food source for wasps. The entrance design of most beehives does not help to discourage their feeding habits.**

## SOLUTION

For most beekeepers, seeing yellowjackets (*Vespula spp.*) rip into bees at the front of the hive is unnerving, but normally, when compared to other pests and diseases, of little consequence. These insects are using the dead and dying bees at the entrance and on the landing board as a food source for protein. In the United States, it takes until midsummer for the populations to reach large enough numbers for beekeepers and homeowners to begin to notice them. They remain active until well into the fall season, and, depending on the climate, can even overwinter to form large nests. In some species, multiple queens add to the wasp population growth. If the climate is warm and winters are mild, these large nests can pose a more serious threat to bee colonies than wasps wintering in colder climates.

This behavior can become a more worrying issue when raiding yellowjackets are emboldened to actually begin to enter the honey bee colony, where they will be attacked by colony defenders. If a colony is weak or small, as in the case of a nucleus hive, bees will have much more trouble defending themselves. In addition to eating adult bees, in some instances wasps will rob honey stores and disturb the wintering colony. The only real solution is to relocate the colony entrance or reduce it.

# 68 Animals are visiting the apiary at night

## CAUSE

The brood and honey stores in bee colonies have always been in demand by any animal that can gain access to the larder. The reward for the thief is high quality food that comes at a painful price. In particular, vermin and birds desire the dead and dying bees that accumulate at the hive entrance.

## SOLUTION

This issue has particular emotional ramifications for many beekeepers. They want to keep their bees healthy and productive, but wildlife may continually harass the colony. It is not uncommon for such colonies to be cranky during daylight hours. Animal excrement, packed with bee parts, may litter the area, or if a skunk has visited, the area may still have some of the scent associated with that animal. Muddy prints may remain on the hive front or on the landing board.

Raising the entrance higher from the ground or improvising various barriers may deter animals, such as rolled 8in (20cm) diameter chicken wire placed at the entrance. Animals do not like standing on an unstable structure and the wire allows defensive bees to sting their underside.

For those individuals prepared to go to greater extremes, live trapping may be considered. All local ordinances and trapping regulations must be followed. Due to concerns about rabies in the United States, mammals cannot be routinely trapped. Clearly, if use of a firearm to control animals is chosen, all relevant regulations must be obeyed.

For most beekeepers, a more acceptable approach is fencing or even electric fencing. The initial expense is greater, but once an effective barrier has been established, future vermin visitors will be excluded.

# 69 The brood smells foul and has punctured cappings

## CAUSE

A sour smell and ragged punctures in brood cappings are common characteristics of American foulbrood. It is caused by a bacterium, *Paenibacillus larvae*, that produces spores that are tough and can remain viable for many decades.

## SOLUTION

American foulbrood (AFB) disease is feared and respected by most beekeepers—and it should be. A colony with more than a few cells of AFB has a poor chance of recovery. The traditional solution is to destroy the infected hive and the bees. In this radical way, the remainder of the colonies in the yard are protected from exposure caused by robbing and drifting bees. Not surprising, beekeepers are frequently reluctant to employ this radical procedure, but if all combs and contaminated honey are not completely destroyed, healthy bees in neighboring colonies are put at risk.

Where legal, antibiotics can be used to eliminate the vegetative stages of American foulbrood. Following the label rates for quantities, antibiotics can be applied as a powdered sugar/antibiotic mix on outer edges of brood frames. Generally, it should be applied three times at three-day intervals and will suppress the increased development of the disease. However, when the treatment ends, AFB will commonly re-occur.

Another traditional technique requires shaking all the bees from infected combs onto new frames and wax foundation. The bees are forced to rebuild the combs and without brood to feed, it is thought that their feeding glands are purged. Again, if antibiotics, such as Terramycin or Tylosin are permitted, these broodless bees can also be treated.

✳ *Beneath ultraviolet light, American foulbrood scales will fluoresce and show up as light blue. The infected equipment can be unknowingly stored away—sometimes for many years—possibly to be sold cheaply at auction. The combs, much more than the wood, will still contain spores that have the potential to once again blossom into a new AFB infection. If the history of old equipment is unknown, be careful when purchasing it.*

## DOUBLE-EDGED SWORD

One the one hand, antibiotic treatments have kept susceptible honey bee strains alive. Had beekeepers not been using antibiotics for several past decades, AFB would probably only affect about 2–3 percent of colonies. However, the use of antibiotics has reduced the incidence of the AFB. Without them, commercial beekeeping would have been much more problematic.

# 70 Wax worms are destroying the combs

## CAUSE

The greater wax moth (*Galleria mellonella*) and the lesser wax moth (*Achroia grisella*) are two premier lepidopteran bee comb degraders. The damage occurs during the larval stage as the larva tunnels through combs in search of protein from pollen and cocoons predominantly in the brood nest. An unusable morass of silk webbing and frass remains in the comb remnants.

## SOLUTION

One of the unintended aspects of modern apiculture is the development of food resources for wax moths. A beekeeper's storage building, with stacks of hive bodies filled with combs, provides massive food resources for adult wax moths. Comb destruction is severe and cleanup is laborious.

Paradichlorobenzene (PDB) has been used for many years as a comb fumigant. Boxes are stacked 5–6 feet (1.5–1.8cm) high, the bottom is closed off, and all joints are taped. Six ounces of PDB crystals are placed on a paper pad on frames at the top of the stack, and an outer cover is placed over this. The crystals need to be replaced every six weeks.

Wax moths do not tolerate sustained coldness well. Equipment can be frozen to eliminate the moth. Since this insect causes extensive damage to stored combs or to combs in weak colonies, other fumigants and procedures have been reported to be useful. Even so, combs—especially brood combs—cannot be left unprotected.

In the United States, the wax moth cannot overwinter in the northern tier of states. Depending on the season, the moth regains lost territory each year only to lose it again as the weather cools. Beekeepers in warm climates must deal with this pest year round.

## GALLERIASIS

As the wax moth lava tunnels through combs, it frequently tunnels through developing bees. If a bee is not killed by the tunneling event, it may very well be constrained by the moth's silk tunnel. These trapped bees will uncap themselves, but not be able to leave the comb. This condition is called Galleriasis and is frequently found in weak or diseased colonies.

❋ *Wax moth tunneling and webbing has made this frame unusable. If this frame was given to a strong colony, the bees could completely remove any remaining wax and webbing and build natural combs. Many of the cocoons would be left in place but they would be covered with propolis.*

## CHAPTER SEVEN
# POLLEN AND POLLINATION

The relationship between bees and the plant blossoms they pollinate is biologically beautiful. After all the years of scientific and casual observations on the plant/bee relationship, there are continually new discoveries and scientific advances. Around the world, most ecosystems depend heavily on pollinating insects in innumerable ways. As the human population has increased, agriculture continues to evolve to provide required food. Increasingly, our honey bees exist in an environment that is foreign to them. Beehives with removable frames—no matter what the design—are not natural nests for honey bees. Yet to meet our pollination needs, modern beekeepers must continue to experiment and develop methods to manage artificial populations of bees, especially honey bees. These supplemental bee populations are progressively more important in the large-scale production of our world's food supply. Yet, at the bee/plant level, the goal is simply to acquire food resources and transfer pollen.

Many of the problems in this chapter are at the advanced end of beekeeping practice, such as the difficulties involved in moving large numbers of colonies to growers' pollination sites, or the trapping and storing of pollen. Others are issues that all responsible beekeepers should be aware of, such as the need to avoid pollinating weed crops, and the more wide-reaching problem of reduced natural bee forage.

# Too much pollen is packed in the brood nest

## CAUSE

**If a colony is populous and the open brood population is low, extra stored pollen will accumulate in and around the brood nest. If a healthy colony is crowded, pollen will be packed into a crowded brood nest.**

## SOLUTION

This is a seasonal issue rather than a specific colony problem. Pollen flows are seasonal and weather dependent and sometimes last only for a few days. On those days when pollen is available, foragers make a dash for it. If a colony's brood nest is crowded, foragers will increasingly convert brood production areas into pollen and nectar storage areas. At some point, so much is stored in the brood area that brood production is hampered. If left unaddressed, the crowded brood nest will likely cause swarming behavior to start.

You should try to keep the centralized brood nest in balance. Brood frames near the center should primarily contain maturing bees encircled by an obvious band of pollen, with some honey stores near the comb top, in the corners, and along the bottom. During intense flows of pollen and nectar, watch for brood frames that are too full of either pollen or nectar. Over time you should learn when to expect these seasonal high flows and provide extra empty brood boxes before needed.

A possible solution to the crowded nest syndrome is to remove full frames of pollen and freeze them. They can be given back to the colony during times when pollen is not readily available. Queen producers often use stored pollen frames to help produce robust queens. Pollen traps can be used to gather pollen for use later during the season. Lastly, honey from crowded hives can be held back to use in making splits later in the season. Nothing is wasted.

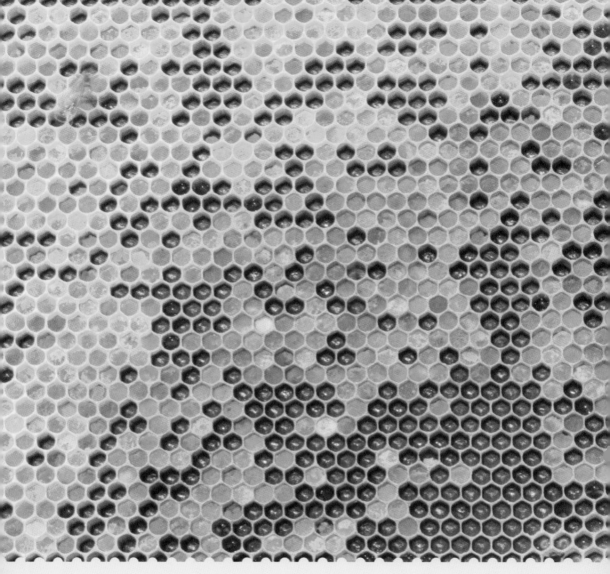

## POLLEN FIX

Pollen has a shorter storage life than honey and is only viable for barely a year. However, even in this depleted state, it still has attractive qualities for protein-hungry bees. If pollen stores become depleted, the band of stored pollen surrounding the brood will be exhausted. Providing a pollen supplement can help colonies that are experiencing a pollen shortage condition.

❋ *Pollen-hoarding behavior has a genetic basis. Research results have shown that high pollen-hoarding strains of bees can be selected and will hoard pollen to such an extent that practically no brood can be produced. In the photo, the only brood present is in the lower right-hand corner.*

# 72 There are drones stuck in the pollen trap

## CAUSE

**Pollen traps are designed to remove the pollen from worker bees as they enter the hive. As drones are larger than workers, they cannot squeeze through the worker-sized openings in pollen traps. Since drones wander from colony to colony, they may accumulate in hives or at entrances.**

## SOLUTION

This restricted drone movement is an unintended consequence of pollen trapping. When pollen traps are in place, monitor the drone activity; if they are jamming entrances or exits, remove either living or dead drones. Nearly all trap designs have drone escape devices or ports that work so long as the entrance does not become blocked by dead drones.

Increasingly, drone brood combs are used as bait to lure Varroa mites into drone cells. After the cells are capped, the drones, along with the mites that are in the cells with them, are frozen. While this is a currently accepted procedure for Varroa population suppression (see Problem 62), drone combs shouldn't be left unattended during times when pollen traps are in place. Drones in colonies with a notable incidence of Varroa infestation may have non-functional wings. If the pollen trap is in place for more than five to seven days, it is important to open the hive and expose the pollen trap to make certain that drones are not accumulating on the pollen-collecting grid. (The pollen trap is underneath the lowermost hive body.)

Though unrelated specifically to drones, while the pollen trap is exposed, the collected pollen should be checked for the presence of wax moths and small hive beetles. They are both pests of stored pollen.

Lastly, the beekeeper should not have a pollen trap on the hive if queens are trying to take mating flights.

# 73 The bees are pollinating weeds

## CAUSE

It's all in a good day's work for insect pollinators. If the blossoms offer a beneficial reward and the flower morphology is compatible with the bees' anatomical structures, they will readily visit plants that are deemed noxious or invasive by human standards.

## SOLUTION

Both the beekeeper and the honey bees are somewhat trapped with this issue. Honey bee food sources are decreasing and surplus honey crops are declining. Even if they try, beekeepers cannot really prevent bees from visiting undesirable plants. Neither the bee or the beekeeper is accountable; however, care should always be taken to avoid the conscious spread of these plants.

In some cases, such as sweet clover (both yellow and white, *Melilotus officinalis* and *M. alba*), Purple loosestrife (*Lythrum salicaria*), and Viper's Bugloss (*Echium vulgare*), these are all excellent honey producers and may even have some beneficial attributes, but they nonetheless also have characteristics that are highly undesirable.

Spirited discussions have sometimes occurred when various honey plant lists present flowering dates and plant attributes of undesirable plants. Though bees and beekeepers may at times profit from the honey crops that come from these undesirable sources, these plants should never be propagated and dispersed. The list of invasive plants is long and detailed, and beekeepers should always be sensitive and informed.

# 74

# The pollen I have collected is decaying

## CAUSE

Honey bee collected pollen is bacterially active. If pollen is not properly stored in combs by house bees and high humidity conditions are at hand, the pollen stores could begin to mold. This spoilage generally occurs in a weak colony. Pollen collected from pollen traps that is not dried is also prone to support mold and fungal growth.

## SOLUTION

When foragers in the field collect pollen, individual pollen grains are made sticky with regurgitated nectar or honey from the forager's crop. They are gathered from the bee's body and formed into a pollen pellet. In the hive, these pellets are mixed with honey into a mixture called "bee bread." Under the bees' management, these pollen stores are protected from mold and fungal growth.

When a pollen trap is used to capture incoming pollen loads from foragers, the brightly colored pollen pellets are still moist and damp. You will need to clean detritus from the collected pellets and lightly dry it under a heat lamp or in an oven. Properly dried pollen pellets will have lost the softness they had before the exposure to heat. If a common home freezer is used for long-term storage, dried pollen should be stored in 8oz plastic bags for freezing. Surprisingly, if larger quantities are frozen, even at freezing temperatures, bacterial hot spots will develop in the center of the frozen pollen and degradation will occur. Commercial freezers routinely operating at lower temperatures will prevent this decay from occurring.

Even if the pollen will be quickly used, it should probably be frozen to destroy any early stage wax moths eggs or larvae that may be present in the stored pollen. Once dried, frozen, and then removed for use, it must be placed in wax-moth-proof containers.

## TASTY TREAT

While pollen is used by some people as a food product, most dried pollen pellets are re-fed to bees for brood buildup. Just a small amount of natural pollen in bee protein mixes makes the mixture more attractive to the bees. When using dried pollen, there is a risk of spreading bee diseases. It's probably best to collect your own pollen to feed to your own bees.

❋ *Collecting pollen is a specialty crop for some beekeepers. Several styles of traps are available through commercial beekeeping equipment suppliers. Depending on the quality of the pollen product, collected pollen can be marketed to health-food markets.*

# 75 My pollination colonies are mixed sizes

## CAUSE

Individual queen productivity varies so colonies grow at different rates. Some colonies will need space more quickly than others. When relocating colonies, this mismatched equipment will result in an uneven load of hives that will be difficult to transport and the hives will contain bee colonies of uneven populations. This could confuse the hive rent payment process and cause more work when loading and unloading.

## SOLUTION

The best approach is to standardize the physical size of the colony as well as the population within the hives. This is not a bad idea even for the beekeeper who is moving just a small number of hives. Having colonies in hives that are all the same height makes loading and stacking colonies much easier. Hives with lightweight bottom boards and simple flat board tops are the cheapest and lightest. Colonies to be used for supplemental pollination are the not the same as colonies that are to be used for honey production. Commercial pollination colonies are generally smaller.

For pollination colonies, you should standardize queen management so that all queens are installed and replaced at the same time. Again, this would be a proper management procedure even for colonies that are never to be moved. As pollination season nears, the colonies' population should be equalized. Aim to have similar populations of bees covering similar brood populations in all hives. Standardizing hives will make payment negotiations clearer for both parties.

To acquire the bee populations needed, syrup and supplement protein should be abundantly supplied. An aggressive approach to Varroa control will have to be started and maintained. These bees will be worked hard and payment is being charged for the bees' labor, so the bees must be as healthy as possible.

## AT SHORT NOTICE

The beekeeper needs dependable vehicles that can get into wet areas with short notice. Commercial growers are commonly on a tight pesticide application schedule, and the beekeeper needs to be ready to move hives in and out on short notice. Bad weather is rarely an excuse.

❋ *For mechanical loading and unloading, commercial hives in the United States are usually in two deep Langstroth hive bodies. The beekeeper hauls the bees to the grower's location and uses the grower's loader and farm wagon to put the bees in location.*

# 76 Too few bees are pollinating the target crop

## CAUSE

Once the available blossoms in a target crop have been visited, foragers will begin to forage beyond the desired crop boundaries.

## SOLUTION

Growers who express this concern are well within their rights. They are paying a meaningful rental price, and in some instances, the targeted crops are only blooming for a short time. They want their crops set. The foraging bees have their own agenda. As quickly as possible, the foraging population will partition the available resources. In this way, enough bees will remain in an area to visit the available blossoms while other foragers search farther afield for new, less densely pollinated areas. Foraging bees compete not only with other foragers from their hives, but also with foragers from other colonies and even other species of pollinating insects. If the field is large enough, many pollinators will just work along the edges of the field while the center of the planting remains unexploited.

If possible, pollinating colonies of bees can be set in "pollinating islands" near the center of the planting to offset this natural behavior. While pollination is paramount during blooming times, consideration must also be directed to field workers and possibly irrigation equipment. Accessibility and soft soil may also limit where colonies can be located for pollination services. In recent years, various honey bee attractants have been developed that are intended to lure bees to target crops and ignore the food rewards of competing crops. To some degree, at the outset, these products can be said to work; however, foragers will soon learn that the reward for responding to the attracting odor is a bit bogus and move onto the undesirable plants.

✿ *Once the blossom has begun to age, the window for successful pollination quickly closes. The fruits that develop will comprise the crop that the grower must market. Low quality products will result in reduced income. The inadequately pollinated cucumber shown in the photo has limited value and will probably be discarded. It becomes a challenge for the grower to get just the right number of foragers at just the right time.*

# The bees won't eat certain types of pollen

## CAUSE

Not all nectars and pollens are beneficial foods for honey bees. Producing pollen and nectar is a costly endeavor for plants and not all pollinating insects are invited to visit the blossoms of these plants.

## SOLUTION

Some of the commonly cited reasons why plants produce toxic pollen and nectar products are: to restrict microbial growth, to protect against herbivores, and, of course, to select for specific pollinators rather than obliging all pollinators. There are several types of sugars that are not beneficial to honey bee development, for example lactose, melezitose, and raffinose.

A plant's production of nectar and pollen is a complex physiological procedure and is affected by many conditions that are beyond the beekeeper's control. Some examples of some of the factors that affect nectar and pollen production are: flower position on the plant, age of a specific flower, plant variety and species, amount of available sunlight, relative humidity, and daytime and nighttime temperatures.

If the sources truly are naturally occurring and if it can be shown that these specific plant products are harmful to honey bee colonies, the only real option is to relocate the colonies out of the range of these plants. In reality, it is most difficult for the backyard beekeeper to tell if the food sources in the area are toxic. Keep in mind that the blooming season is commonly only a few days, so the situation will likely resolve itself.

✳ *If exposed to natural toxins, bees are not without some instinctual recourse. In the photo above, for unknown reasons, worker bees have entombed about seven pollen cells with a propolis and wax covering. Upon opening a cell, it contains what appears to be oily, dark-reddish pollen that the bees apparently would not consume even to remove it from the colony. Seemingly, bees do avoid some food sources that would appear to be perfectly appropriate for consumption.*

# 78 The bees won't consume pollen supplements

## CAUSE

Pollen substitutes are unattractive to honey bee pollen foragers. Though the nutritional components may be present in quantities sufficient to meet their protein needs, the product may not have the necessary attractants. Additionally, an ongoing natural pollen flow will diminish the forager's willingness to consume artificial sources of food.

## SOLUTION

As long as the bees are gathering nutritionally balanced pollen, it isn't a true problem if they ignore the artificial protein sources. In recent years, commercial companies have developed a variety of pollen supplements. Their individual acceptance to the bees will vary between colonies and the annual season. The same colony will consume different rates during different seasons.

Bees within a colony can be forced to eat pollen supplements by positioning it in the center of the brood nest. Bees have a tendency to remove this obstruction from the nest. Some will be eaten but, no doubt, some of the product will be discarded. When feeding pollen substitutes, monitor the colony for the presence of small hive beetle (SHB). They are attracted to pollen supplements.

Natural pollen can be blended with powdered pollen supplement to make it more attractive to the bees, but beware: diseases can be spread in this way. It would be best to collect your own pollen for this purpose.

Some supplements are fed as a powder, while others are fed in paste form. Occasionally, products intended to provide trace elements, minerals, and salts are offered by the beekeeper. Such products are much like humans taking vitamins. They may or may not be needed. It is important that some pollen supplement be provided to the colonies—even if they don't immediately accept it.

❋ *When natural pollen is in abundance, bees will be reluctant to consume artificial pollen supplements.*

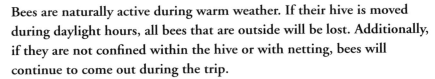

# 79 Many bees were lost during a hive move

## CAUSE

**Bees are naturally active during warm weather. If their hive is moved during daylight hours, all bees that are outside will be lost. Additionally, if they are not confined within the hive or with netting, bees will continue to come out during the trip.**

## SOLUTION

In order to prevent losing the foraging field force that is outside the hive during the day, it is best to move hives into and out of a pollination site after dark. This procedure works best when moving only a few colonies (maybe 10–50 colonies) to provide bees to only one commercial grower per evening. Make all preparations during daylight hours. Using nylon ratchet straps, colony components should be soundly strapped together. Because entrances are generally left open, you will need to be fully suited and gloved. If working alone, a hand truck or some other type of cart should be used when loading the hives (see Problem 22).

Another possible solution is to move colonies during the day, but leave a few colonies at the original site to provide hives for the returning foragers that are otherwise hiveless. This will require a second trip to retrieve these remaining colonies.

Large commercial beekeeping operations use different techniques for moving hives than beekeepers with smaller loads. Regardless of the moving technique used by commercial keepers, significant numbers of bees are lost. For a commercial pollination provider, this loss may not be a bad one. If the capped brood population is high and the queen is young and prolific, the older bees lost outside the hive are replaced within a few days with younger, more vibrant foragers.

## ON THE MOVE

Before moving bees to a new location, the apiarist should make preliminary daylight trips to the pollination site to become familiar with it. When returning to pick up the bees in a few weeks' time, vegetative conditions may look much different, and colonies can be overlooked or difficult to find. Trucks and trailers need to be mechanically sound and dependable.

❋ *Before relocating hives, heavy supers of honey like this one should be removed to make the hive lighter and less top heavy.*

# The bees are struggling to find good forage

## CAUSE

**Widespread herbicide use, the development of efficient mower technology, and the evolution of modern agriculture are some of the reasons that have resulted in the reduction of bee forage.**

## SOLUTION

For many years, bees and butterflies were plentiful. Their populations were a naturally occurring phenomenon and few people gave serious thought to the possibility that these beneficial insect populations could decline. But they have now reached unprecedented low levels, which is a cause for concern not just for beekeepers and agriculturists, but to all of us, since so much of the food we live off relies on pollination by bees.

You can plant clover and other bee food plants on your property, and also encourage neighbors and civic groups to do the same. But bees need a large foraging area to find all the food components they need. The bee habitat needs to be a mixture of plants that provide forage sources throughout the warm months. Even if the plants are copious nectar and pollen producers, bees cannot survive on a few species of flowering plants.

Ultimately, beekeepers need to work together to address this global issue and initiate change for the better. Get active in the local bee club to raise awareness concerning the decline in flowering plants. Master Gardener groups also make good partners in programs that support pollinators and butterflies. There are new restoration programs that are intended to encourage landowners to provide bee-friendly habitats. Familiarize yourself with these programs and support the cause. If large numbers of landowners did just a small planting, overall foraging conditions would improve.

*❊ The decline in flowering plants has led to a reduction in bee forage*

# FLIGHT TIME

Food source location plays an important role. The farther that bees must fly, the more expensive it is for the forager. The average bee has been estimated to only have about a 140-mile (225km) flight lifetime with a maximum of 500 miles (800km). While it is true that bees can forage as far as five miles from the food source, these distances push the limits of efficiency and productivity within the colony.

# CHAPTER EIGHT
# PRODUCING AND PROCESSING HONEY

Since beekeeping's earliest years, honey has been used as a sweet, nutritious food and as a medicine. During the past decade, in most countries small-time honey producers have enjoyed increasing success. Recent uses in distilled spirits and sauces have only added to the demand for both locally and commercially produced honey.

Making honey is what bees do best, but there are certain pitfalls the conscientious beekeeper should watch out for in order to maximise the quantity and quality of their honey crop. From bees refusing to leave the honey supers to honey granulating when it is still in the comb, there are a number of problems that can occur before the honey has even been extracted.

The extraction process itself will vary, depending on the size of your operation, but many of the challenges to be faced here are common to all beekeepers: clogged honey filters, bees flocking to the extraction site, and accidental spillages, to name a few.

There is little more satisfying to a beekeeper than seeing jars of their bees' own honey lined up, labeled, and ready to go on sale or be shared with fortunate friends, family, and neighbors. The liquid gold really does make all the hard work worth it.

# 81 The bees won't leave full honey supers

## CAUSE

**When honey supers are full and nights are warm, bees have no particular reason to leave them. However, if a fall crop is produced and night temperatures drop, bees will more readily move down to brood areas.**

## SOLUTION

If bees need to be removed from honey supers during warm months, devices like bee-escapes that fit in the inner cover handhold provide a one-way exit for forager bees to leave top supers but not return. When removing honey in the spring, these devices will probably not work well. Bees need an impetus to leave these supers; otherwise, they will remain there for the night. The next day, bees will begin to trickle from these supers, but several days may be required for all to leave. During that time, problems can arise. Since the bees can no longer return to the higher supers from the inside, robber bees may find a rotted corner or crack in a honey super and begin to rob the colony of its unprotected honey.

Commercial beekeepers commonly use fume boards to incite bees to leave the supers in commercial apiaries. Bee supply companies provide various bee repellants, but they may not be available in all countries. Normally, some types of these products are available to beekeepers with smaller hive numbers. These bee repellants all work the same way in that they aromatically push the bees from the honey supers and can have a noticeable odor.

Of course, simple bee brushes work okay, but brushes can really rile the bees into a defensive mode. Various styles of bee blowers or shop vacuums also work well. Specific models can be purchased from bee supply companies. Alternatively, a leaf blower will also work.

## DRAWN TO LIGHT

The fact is that no device or procedure removes 100 percent of the bees all the time. In all cases, some bees are left in the supers. They are attracted to extracting-room lights and can drop into open tanks of honey. Entrance traps can be improvised to encourage bees to move to the outside where a bait hive can be set up to accept them. At night, lights can be used to lure them outside.

❋ *Full honey frames like this one are more easily freed of defending bees than combs that have open cells remaining. The few bees on this honeycomb can be easily brushed away.*

# 82 Some honeycombs do not have wax cappings

## CAUSE

Beeswax production is closely related to the vigorousness of the nectar flow. If the nectar flow slows or even stops, wax production in the colony will follow. Bees are reluctant to consume stored honey to be converted to beeswax. If the flow should begin again, wax production by bees in the colony will also start again.

## SOLUTION

Heavy syrup can be fed to the colony to provide a simulated nectar flow. If many colonies are involved, this can be a laborious solution that verges on impractical; but if only few colonies need help, this is a good way to go.

While nearly any feeder can be used, top feeders with a large storage capacity are probably best. It is hard to estimate how much syrup is required, but if the colony is populous and healthy, and if there is a super or two needing wax for cappings, it can take several gallons of heavy syrup. Heavy syrup is a relative term but is typically a mixture of two parts sugar to one part water. This amount of sugar will require some heat to drive it into solution. Alternatively, light sugar syrup will probably be closer to a 1:1 ratio of sugar and water and is used as a spring stimulant to entice bees to begin the season. This mixture will also require heat, but hot water from the tap will most likely be adequate.

If the uncapped honey is only a small part of the surplus crop, many times it can simply be extracted and blended with normal capped honey. Although a bit costly, a refractometer is helpful for determining moisture content. If the uncapped crop is very near the minimal moisture level (18.6 percent moisture), it will have little effect on the normally capped crop.

## FROM NECTAR TO HONEY

If the crop is thin, watery, and drips from the combs, at this stage it is still very much nectar and should not be extracted. Most likely, the bees will continue to passively process the crop by removing moisture from it until it is nearer the moisture content of honey, but the bees will need a nectar flow to begin to generate wax for the cappings.

❋ *Pictured is a frame of partially processed nectar. A few days of rain disrupted the nectar flow, resulting in the bees slowing the capping process. In this instance, the weather improved and the capping process ended successfully.*

# 83 The extracting room is cramped and inefficient

## CAUSE

**Every processing area is unique to the individual beekeeper. As colony production increases, the extracting area can become crowded and inefficient.**

## SOLUTION

Every beekeeper who keeps bees for honey production will have to address the issue of where and how the surplus honey crop will be processed. As the beekeeping operation grows in size, so does the extraction operation. It is common for new beekeepers with only a few hives to set up an extracting operation in the kitchen or a similar room. Honey processing is inherently messy. Due to this clutter, few beekeepers remain in a kitchen extracting room for many seasons.

When you are ready to upgrade to a larger extracting space, make sure you bear the following in mind. Electrical power is obvious. In fact, higher voltage power is usually desirable, especially in the United States. An abundant supply of very hot water is useful for cleanup. The room or facility should be well lighted. Because wax and propolis are difficult to remove from rough cement, firm washable floors are necessary. Epoxy paint is a common coating for walls and floors. The room needs to be reasonably bee-proof to prevent robber bees from entering. For larger operations, a loading dock is desirable; when possible, wide doorways make it easier to maneuver carts of heavy honey supers inside.

Many beekeepers routinely process their crops in areas that have few of the attributes listed above. Even so, it is always important to remember that even though honey is a hardy product, it still needs to be processed as a human food crop, with attention to relevant health and safety and food hygiene regulations.

## POP-UP EXTRACTION UNIT

A dedicated, freestanding building in which to process your honey crop is the ideal, but the construction of such a building might not be affordable or practical for all beekeepers. You could consider modifying an existing space such as a two-car garage for a couple of months to process significant honey crops. Afterward, the entire extracting setup can be broken down and stored away until the next season. In this way, income is not required to construct and maintain buildings.

❋ *In order to extract honey from the comb, the beekeeper is removing the wax cappings. After uncapping, a honey centrifuge will be used to remove the honey from the combs.*

# 84 Many bees are coming to the extracting area

## CAUSE

Honey crops are usually extracted after all nectar flows have ended. During summer months, strong colonies have many unemployed foragers that have little to do. The aroma of honey lures these bees to the extracting area in large numbers, where they quickly become pests.

## SOLUTION

Ideally, honey crops should be extracted behind screened enclosures or within a room with bee-tight doors and windows. It is tempting to extract outside, where the mess made by dripping honey and wax won't be of concern. However, the scent of the honey will be an odor beacon to bees. Even if extracting behind bee-tight screens, you should anticipate significant numbers of bees flying around the outside area, which could be disconcerting to others.

If there is simply no place to perform the extracting operation except outside, then consider doing it during early evening and into the night, when foragers are not active. There will still be a few bees that were inadvertently left in the honey supers and they will fly to nearby lights; therefore, the extracting area should not be directly beneath these lights.

If practical, the beekeeper should try to keep the bees that have found the source confined to the area. If they return to the colony, they will only recruit more foragers to the area.

## FEEDING FRENZY

The chaos that can result when bees have free and open access to exposed honey can be impressive and the number of bees killed can become significant. Bees first on the scene will begin to imbibe honey and while they are drinking, and as the crowd grows, other bees will begin to stand on those first bees until they are pushed under. Ultimately, thousands of bees can be killed in this frantic cycle.

✽ *The open honey in the extractor tank and the honey odor being fanned by the rotating extractor-basket air make outdoor extraction a bad idea. In bee-populated areas, tens of thousands of bees could come searching for the honey source. Using smoke on these excited bees is useless.*

# 85 Honey has granulated in the comb

## CAUSE

Some plants, such as the mustard family (genus *Brassica*) produce a good nectar crop that rapidly granulates even before the nectar flow has ended. Beekeepers whose bees produce a honey crop from such plants must extract the crop almost as fast as bees store it.

## SOLUTION

There are few solutions for re-liquefying honey that has granulated in combs. New beekeepers may be surprised to learn that this situation is reasonably common in bee colonies. Even when extracted and stored in containers, it is widely known that, given time, most honeys will granulate. This is a normal chemical process. Honey is a super-saturated solution of sugars. The excess sugars are precipitating and a more normal concentration level is being chemically sought. While this is a normal process, fermentation does sometimes occur in the watery band of honey that forms at the top of granulated honey. This indicates that the honey has not been high-heat processed.

When parts of honeycombs are granulated, the extracting process for those parts of the comb that still contain liquid honey proceeds in the routine fashion. After uncapping, some of the granulated crop may be expelled from the combs and become blended with the extracted honey from other plant varieties. This "sugared honey" probably serves as granulation seed for the solidification of the remainder of the extracted crop. Since most extracted honey will granulate anyway, this is not thought to be an urgent situation.

While extracted honey can be re-liquefied with gentle, low heat, honey granulated in the combs is essentially nothing more than bee feed. Just as bees can consume granulated table sugar, they can readily consume granulated honey from the combs. Little else can be done with it.

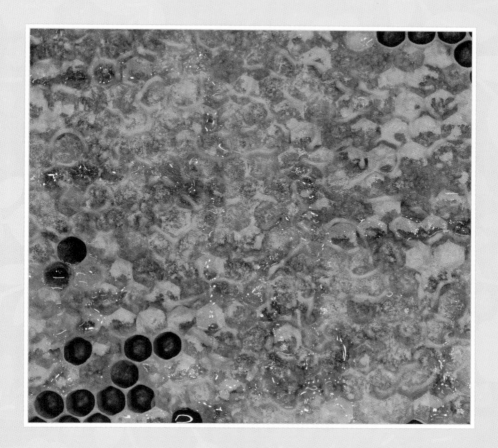

## NATURALLY GRANULATED

Granulated honey is not an "all bad" situation. Honey that naturally granulates often has a gritty or sandy texture that is something less than pleasant. Creamed honey products are nothing more than finely granulated honey. Nothing else is added. Commercial honey packers expect drums of honey to be granulated. Large ovens are used to re-liquefy these blocks of granulated honey. One advantage is that if punctured, drums will not leak badly.

❋ *Honey that has granulated in the comb takes on a slightly cloudy look. It is not always full frames that are affected, but possibly only parts of the combs. Bees will eat this honey. If the granulated honey is slightly fermented, bees will still consume it.*

# 86 My comb honey crop is disappointingly small

## CAUSE

**Bees do not readily accept comb honey equipment. No doubt the morass formed by the various walls and restrictions that make up the comb honey sections are not the bees' perception of a perfect hive.**

## SOLUTION

Producing perfect comb honey products is nearly an art form. For many years, various sizes of thin-walled basswood boxes were manufactured and bees were tasked to fill them with capped honey. Sensible bees were rarely eager to fill these small boxes. Various techniques were thought to be helpful when trying to produce this delectable product.

All of these comb boxes had one feature—corners. Bees are notorious for leaving small openings in corners as passageways, so many comb honey sections had open corners. After years of using basswood boxes, circular plastic sections were developed that obviously had no corners. Along with preformed plastic clamshell devices, these round sections are still marketed today. Yet, the bees' reluctance to work even the modern section devices is not greatly improved over the old wood box sections.

Preparations must be made to produce comb honey, and the most important preparation is to use thin foundation or no foundation at all. Since comb honey is intended to be eaten, there cannot be a thick piece of foundation in the center. Secondly, many bees crowded into a full brood nest with only empty honey supers for storage is frequently recommended. Therefore, swarming is a common issue. If a large swarm can be captured and installed onto mostly full brood frames with empty comb honey supers on top, there is a good chance a comb honey crop can be obtained; however, there must also be a good ongoing nectar flow.

## MAKING COMB HONEY

Unfinished sections are common in comb honey production. In many instances, the beekeeper and his family and friends consume these imperfect comb sections. All comb honey crops have some of these imperfections. Near the end of the flow, combine these sections and add a feeder to provide syrup to entice bees to finish the incomplete sections. Feed only enough to encourage production of cappings and not honey storage.

❊ *Chunk honey is a type of honey most commonly produced in the southeastern United States. A piece of comb honey is put into a jar and the surrounding empty area in the jar is filled with extracted honey. The wax combs and honey are used as a spread on bread and biscuits. The thin wax comb can be safely eaten.*

# 87 The honey filter keeps getting clogged

## CAUSE

Freshly extracted honey can have a lot of heavy natural flotsam that will readily clog simple filters. Beekeepers with small-scale equipment must deal with constantly changing or otherwise maintaining filters during the extracting process.

## SOLUTION

The filter is an integral part of the extracting line, and potentially one that is frequently annoying. It is messy to maintain and often causes spills and overflows. For the small-scale beekeeper, there are few commonly available honey filters that are not maintenance-intensive.

The core problem is that most people, including some beekeepers, are repulsed by the reality that bees truly are integral to the production of honey. The only way they have to manipulate it is by filling their internal honey crop and regurgitating it wherever it is needed. Because it reminds consumers of where the honey product actually originated, an extraneous bee leg in a honey jar is offensive. Consequently, the honey filter is important in removing these course natural contaminants.

Bee manufacturing companies offer a variety of sieves and filters that capture these materials, but invariably, these devices clog as they are designed to do. In past years, cheesecloth was used as a filter, but this material added significant amounts of lint to the honey product. This was a cosmetic contaminant, and honey show judges downgraded the product because the lint was not a natural colony component.

Heavy-duty airless paint filters have been successfully used to filter both honey and beeswax. The nylon filters are strong and inexpensive. Keeping the filter submerged in extracted honey allows the filter to function much longer. When submerged in honey, large particles do not quickly clog the filter bag wall.

## SETTLING TECHNIQUE

Although it takes a while and some low level heat, there is a simple way to completely avoid the filter mechanism and still get pristine honey. Extract the honey and allow it to sit in settling tanks. Over a 3–4 day period in a warm room or by using heat cables to warm the honey, all the contaminants will rise to the surface and can be skimmed away.

✻ *In this extraction process, honey is being passed through a simple sieve filter to remove any natural contaminants.*

# 88 I'm unsure how to guarantee varietal sources

## CAUSE

Honey is frequently promoted by varietal source, such as orange blossom, blueberry, or basswood. In reality, it is difficult to guarantee what percent of the extracted crop is specifically from the source listed in the name. Normally, it is not an issue, so long as the source is not a popular type of honey such as Melaleuca or Sourwood.

## SOLUTION

Average beekeepers have few ways to conclusively demonstrate that a particular crop was produced from specific plants. The most common technique is to assume that the crop being produced is coming from the primary plant species in the area. Surplus honey crops are primarily from the obvious flowering plants in the area while secondary plants contribute to the honey's delicate nuanced bouquet.

The best you can do is to know the average calendar beginning and ending dates for the typical nectar flow of the desired honey crop and arrive a few days in advance with empty supers and stack the colonies up with empty equipment. If the flow goes well, monitor the progression of the flow. Once the flow has clearly peaked and the blossoms are waning, the supers should be taken off before the next flow, if there is one, starts and contaminates the crop.

But what if your hives are located on the fringes of the nectar flow area? Possibly the desired source is common, but lesser crops are also common. Does the crop still deserve the desired name? Probably, yes—it does. From similar areas there are good and not-so-good varieties of wine, yet they are all from the same area. The consumer is the final judge.

## DNA PROFILING

A large commercial US manufacturer of fruit spreads and jellies sends honey samples to international specialized labs, where the pollen is extracted from individual honey samples and the DNA profile of the honey sample pollen is compared to standard pollen DNA samples of the same nectar-producing plants. The cost per sample is significant but the company is more secure in the source description of the honey.

❀ *Honey from different sources has different flavors and a range of possible colors. The flavor differences are generally slight, but the flavor and color variations between honey from sources such as buckwheat and clover are obvious. Consumers who have been eating honey much of their lives develop a preference for particular flavors. Many beekeepers simply combine all their crops into one generalized honey; however, some beekeepers will specialize in producing honey from specific sources.*

# Liquid honey has accidentally been spilled

## CAUSE

Honey spills and bee stings are inevitable. If one keeps bees long enough, both will happen.

## SOLUTION

The size of the spill delineates the size of the problem. The location of the spill may very well add to the dilemma. Approximately thirty years ago, most honey containers were made of glass. At that time, the situation was always made worse by combining broken glass with spilled honey. Using plastic containers has helped eliminate the glass issue.

Small quantities of spilled honey are little more than annoying, but large spills are challenging to address. Each large spill requires unique consideration. Because sumps and tanks readily overflow, spills are a common issue. Lines from honey pumps bloat and stretch as the honey filter clogs. Heavy drums topple from a loader, and the lid pops off of a drum of 660 pounds (300kg) of liquid honey.

Soap and hot water can be used to clean up small spills, but large spills require essentially any tool or procedure that will work. Flat shovels and squeegees may be required to corral the spilled honey while it is scooped into containers. Wide, flat hand trowels may be useful in lesser spills. Ultimately, after as much of the honey as possible as been removed from the floor, copious amounts of hot water and mopping will bring the floor back to proper standards. Depending on the size and location of the spill, pressure washers may be needed. Feed the reclaimed honey back to the bees or discard it. Salvaging it for human consumption is never acceptable.

❋ *Spilled honey can be*
*tricky to clean up. Any*
*reclaimed honey will not be*
*fit for human consumption,*
*but you may be able to feed*
*it back to the bees.*

# 90 Extracted honey has granulated in jars

## CAUSE

Most varieties of extracted honey, given enough time, will granulate. Granulation is a natural process. Though the honey is still completely safe to eat, many times the customer discards the product, assuming that "it has gone bad." Feeling that they wasted their money the first time, customers are frequently reluctant to buy more.

## SOLUTION

The granulation procedure can be reversed. To prevent pressure buildup, loosen the lid on the jar of granulated honey and place in hot water (uncomfortably hot to the hand, but not boiling). Depending on the size of the jar, the honey will liquefy in about 30 minutes or so. The problem is that once the honey is removed from the heat and put back in the cabinet, once again given time, the granulation process will begin again. Each time the honey is liquefied, some of the delicate aromas and flavors are lost and the honey begins to darken. Heat causes the HydroxyMethylFurfural (HMF) content in honey to increase as the honey is heated and reheated. For the most part this chemical marker is used primarily to identify commercial honey that has been heated to too high a level.

Large quantities of small jars are more tiresome to re-liquefy than bigger, easier to handle ones. If there are a significant number of small jars to liquefy (maybe inventory that did not sell), it is a good idea to either feed it back to bees for conversion to brood or pour honey into larger jars.

For the consumer, the best solution to the granulation issue is not to purchase more honey than you can eat in six months and to thereby keep re-liquefactions to a minimum.

## THE GRANULATION PROCESS

Granulation occurs around fine dextrose crystals (nuclei) that are in the honey. These crystals attract other crystals, and a latticework begins to form that initially causes the honey to begin to lose its brilliant clearness. Depending on the specific chemistry of the granulated honey, the honey increasingly becomes opaque, and finally, becomes a semi-solid mass. When intentionally producing creamed honey products, this stage of such honey is said to be "set."

❉ *All honey will naturally granulate over time. Though it might be a bit harder to get out of the jar, it is still completely safe to eat.*

# 91 My bottled honey is not selling well

## CAUSE

**Some common reasons that honey does not sell well are: being offered to the wrong purchasers, the product has a poorly designed label, an unattractive container, or the honey product is visually unappealing. Competition with other honey vendors in the area could also be contributing to the slow sales.**

## SOLUTION

Honey purchasers and the way they buy their honey products vary from one market type to another. If your honey is stocked by a local store, you should prepare smaller container sizes. Ensure the product is immaculately clean, without any hint of granulation or contamination, and affix a tasteful label that meets all labeling and weight requirements. If possible, make occasional visits to the store to rearrange the honey containers and remove any containers that have begun to look tired. If contact information is clearly presented on the label, you can expect customers who like the honey product to begin calling you directly for larger quantities than those offered at the store.

At farmers' markets and roadside stands, the consumer expects a more rustic product. Regardless, you should keep the containers and the product clean. These customers will probably tolerate a few pieces of wax floating in the honey, but bee parts in the honey product are never appropriate. A more generic label can be on the container, but it should still have clear address information. If your honey stocks are available to supply increased quantities, larger container sizes can be offered. You can also offer a bulk honey tank for refilling customers' containers.

Wherever you are selling your honey, stay abreast of what other honey suppliers are offering and how they are presenting their products.

*❋ Honey should be presented for sale in neat, clearly labelled jars or bottles.*

## GLASS OR PLASTIC?

Glass jars appear cleaner and clearer, but plastic is much cheaper and not inclined to shatter, and is easily recycled. Glass containers can be reused, but usually, they are not. The sterilizing process and the requirement to remove all traces of previous labels is tiresome. Few beekeepers try to reuse glass containers in large quantities. Reusing lids on honey jars for resale is not advisable.

## CHAPTER NINE

# BEESWAX, PROPOLIS, AND SPECIALTY ISSUES

Since beeswax is a side product of the colony that is strongly correlated with the honey production, the only way to get more wax is for the colony to produce more honey. The uses of beeswax are myriad, from candles to cosmetics and beyond, and beekeepers are not the only users of beeswax. Propolis, as a hive product, is omnipresent in every hive, but few beekeepers actively encourage their colonies to produce this unique product. Outside the honey bee industry, most people don't know what the product is or how bees use it.

While most beekeepers focus on honey production, and perhaps pollination services, there are many secondary issues that contribute to the depth and diversity of the beekeeping craft. This chapter focuses on some of the problems that can be encountered when collecting both beeswax and propolis, as well as looking at some of the difficulties associated with the most popular product made from beeswax— candles. Space is also given to some of the unique problems of beekeeping such as how to remove bee droppings from an angry neighbor's car bonnet, and how to deal with that ever present inconvenience of beekeeping— bee stings.

# 92 Beeswax is firmly stuck to plastic foundation

## CAUSE

Wax from plastic foundation inserts cannot be melted off the way it could be in old-style beeswax frames. If rendering wax is attempted, the plastic frames will distort, making them worthless for refurbishing.

## SOLUTION

In general, plastic frames have a three- to five-year lifespan. Techniques for refurbishing these frames for reuse are still evolving. One procedure is to use wax moths to help clean the plastic frames. During the summer months, place frames to be refurbished into an empty hive. Early the next spring, when the moths have degraded most of the wax, the plastic equipment can be pressure-washed. Once clean and dry, dip a small paint roller in molten wax and roll it onto the foundation insert. The clean, wax-coated beeswax frames can then be given back to the bees.

Alternatively, scrape the old wax combs from the foundation insert with a scraper, collect it, and melt it into beeswax cakes. The frame can then be cleaned and recoated with beeswax in the manner described above. You can also use gentle heat from heat guns to soften and remove hard wax from old plastic combs. In general, beekeepers report that the bees, to an acceptable degree, reuse most of the refurbished frames.

If you wish simply to discard the plastic frame equipment, the old comb can be scraped from them for rendering. Although old brood combs are dark and have a lot of propolis in them, commercial wax rendering companies can still reclaim the wax that is there. Many frames will have to be scraped before appreciable amounts of old wax can be accumulated. If approved for use in your area, Paradichlorobenzene (PDB) will probably be needed to prevent wax moths from destroying any collected beeswax. Discarded plastic frames should be recycled using typical recycling procedures.

# Propolis is difficult to gather and process

## CAUSE

Propolis is very sticky when fresh and very hard when dried. After it dries in the hive, it is sometimes called "bee glue." It sticks to hands, tools, and clothing, and is difficult to remove. Bees only produce it during springtime and early parts of the summer.

## SOLUTION

Outside of beekeeping, propolis has multiple uses and value. It has been used in folk medicine for thousands of years. It also has novel uses in human food, in car wax, and as a component of musical instrument finishes. Plastic grids are the most common way to entice bees to produce propolis. Since hive pests can hide in them, bees do not like these open grids in the hive. Essentially, any crack that a bee cannot enter is filled with propolis, thereby restricting the activities of small hive beetles, wax moths, and ants. These plastic grids resemble queen excluders, except that bees cannot get through the grid.

 The grids are inexpensive and can be purchased from all major bee supply companies. They are exactly the size of an inner cover but are much thinner. The grid should be installed on top of the hive immediately beneath the inner cover. A homebuilt alternative is to make multiple ⅛in saw cuts in a ¼in plywood board and lay the board on top of the hive. This slotted board should also be the size of a standard inner cover. The amount of propolis collected will vary from colony to colony.

 If you are interested in collecting propolis scrapings, improvise a rack over a large tub or a barrel half. Hive bodies can be set on the rack and then scrapings can drop into the container below. After equipment has been scraped, add water to the scrapings. Wax and wood scrapings will float and propolis scrapings will sink. The propolis can then be dried and stored.

# 94 I can't get my beeswax candles out of the mold

## CAUSE

**Beeswax is an excellent material for poured or dipped candles, but may cause problems when used to produce molded candles. As the beeswax cools, it shrinks significantly and tends to stick to the mold. Generally, metal molds cause more problems than plastic molds.**

## SOLUTION

Usually, newly poured candles will need to stay in the mold until the next day, allowing time to become fully hardened. Even so, when removed, it is not uncommon for beeswax candles to stick to the mold. When using metal molds, a commercial candle release spray can be used. These products may or may not contain silicone. Plastic molds are treated differently than metal molds. Do not use silicone spray on plastic or acrylic molds. Over time, they will be damaged by its use. There are release products available for this type of mold.

Subjecting the stuck candles to freezing temperatures occasionally breaks the candle loose. Alternatively, warming the mold with comfortably warm water will release some candles. Allow time for the mold to heat and do not use excessively hot water.

A simple procedure of coating the inside of the mold with vegetable oil works well with some molds. This oil can be tried with either metal- or plastic-type molds. Shake out excess oil, leaving only a thin film before the molten wax is poured. Molds can be quirky. Individual testing and experimentation may be required to determine how to best coat or treat a particular mold.

If all goes wrong, be careful not to scratch or scar the mold sides. Do whatever it takes to remove the stuck beeswax. The mold will need to be thoroughly cleaned before trying again. Aggressive cleaning agents may be required.

## A PLEASANT PURSUIT

Producing beeswax candles is a speciality aspect of beekeeping that can be taken to an art form. A well-made beeswax candle with properly sized wicking is nearly smokeless and dripless, and the burning candle produces a pleasant odor. Candle making and candle burning can produce pleasant reminders of the time spent with the bees. However, being a beekeeper is not a requirement; anyone can enjoy the procedure.

❀ *Some molds are a joy to use while others consistently cause difficulty. Heavy, well-made acrylic molds are flexible and are usually good for use with pure beeswax. High quality products typically make the candle production process easier.*

# 95 My pure beeswax candles have a cloudy film

## CAUSE

Much like granulated honey, nothing is wrong with beeswax products that have developed a grayish film bloom over time. This chemical action is an indication that the candles are made of pure beeswax. Most customers do not desire candle-wax bloom, but a select few feel that this characteristic film adds legitimacy and character.

## SOLUTION

If the crystalline bloom reaction is unacceptable, lightly expose the candle to the hot air from a heat gun or from a hairdryer. The bloom will fade back into beeswax, but will return in a few weeks. The hot air procedure works well on molded candles having a design or raised features. If the candle is a simple, smooth candle taper, it can be rubbed with a soft cloth. The friction heat developed by the cloth will also cause the bloom to be reabsorbed into the beeswax. Even so, the reaction will reoccur.

Commercial gloss, provided by various bee supply companies, can suppress bloom development for a much longer time—possibly indefinitely. Initially, the gloss product has a unique smell, but it quickly dissipates, allowing the candle to once again smell of beeswax.

Temperature cycles seem to encourage beeswax to bloom, but it is clear that it takes time for the chemical reaction to occur. If stored above 60°F (15.5°C), bloom will take several months to appear. Beeswax bloom is a natural procedure that visually assures wax purity. True, the appearance to the uninitiated may appear diminished, but if bloom is not present at all and beeswax purity is avowed, let the buyer beware.

❋ *Over time, pure beeswax candles will develop a light grayish crystalline film. This characteristic reassures the buyer that the candle is pure wax that will burn cleanly and without smoke.*

# 96 Bee stings are painful and cause swelling

## CAUSE

**Insect stings, particularly those administered by honey bees, are defensive behaviors intended to protect the species or food resources. When stung, melittin is the primary chemical that is responsible for the pain experienced. Subsequent swelling is the body's effort to flush venom components from the area.**

## SOLUTION

The defensive sting of a honey bee is a normal aspect of keeping bees. Other than some initial discomfort and redness at the sting site, few indicators should remain afterwards. If any bee sting seems abnormal, the affected individual should immediately seek professional medical attention. Putting ice on a sting offers some temporary relief. Over-the-counter medications are also available.

Abrupt movements, perfumes, and exhaled breath are common sting attractants. Even experienced beekeepers should wear a protective veil. Traditional protective clothing is the beekeeper's primary defense. Always have a full set of protective clothing nearby in case it is needed. The smoker is a vital part of beekeeper defense. If full protective clothing is not worn, the smoker should certainly be lit before it is needed and not after the bees have begun their assault.

From a beekeeper's perspective, reasonable pain and swelling at the sting site are normal. However, on rare occasions, sting immunity does not develop correctly. Be on the alert for sting reactions away from the sting site. Rashes or swelling away from the sting site—particularly throat or eye swelling—are indicators that medical attention is needed. Beekeepers need to respect stings and monitor the number of stings being delivered. The number should be small. If a colony stings too much, requeening may produce a more gentle stock.

❋ *The bee's sting is a defensive mechanism designed to protect the species and the colony's precious honey stores.*

# 97 My honey soap is not setting up properly

## CAUSE

In many cases, this happens because the scale is not calibrated correctly, and so measurements and mixtures are not formulated properly. In other instances, temperature may cause a problem: both the fat and the lye should both be at 95°–100° F (35–37.7°C) when mixed.

## SOLUTION

Because it has the ability to attract moisture from the air, honey is a common ingredient in specialty soaps. Soaps made from recipes with a honey component help prevent hair and skin from drying. This moisture-absorbing ability makes honey a common ingredient in soaps, shampoos, and a common component in cooking ingredients.

All soap recipes require measuring and mixing accuracy. Scales and thermometers must be dependable. Increasingly, digital scales are the soap maker's choice. They are precise and inexpensive. Another useful tool is a quick-read digital thermometer. Devices such as these provide a fast, accurate reading.

For novice makers, recipes are available that simplify the process and avoid many of the calculations required that could cause errors and miscalculations. Following the recipe requirements, combine the fat and lye mixture at the required temperature. The challenge is to have both components at the 100°F (37.7°C) temperature at the same time before they are blended. A little more or a little less lye or fat can cause the soap mixture to fail; therefore, do not alter the fat/lye ratio. Exact weights and precise measurements are always required. Even after the soap has set up, it must cure for several weeks before use.

If everything has gone wrong, the mixture can occasionally be saved, but don't depend on that possibility. Be precise and follow recipe instructions.

❋ *This honey soap has been shaped using an attractive mold with a honeycomb and bee motif.*

## TAKE CARE WITH LYE

In soap making, lye must be used and mixed carefully. The primary safety concern is the potential for chemical burns if it comes in contact with human skin. If skin is wet, the burn is even worse. Wear suitable clothing and work in a safe area. Only stainless steel pans and wooden or silicone utensils should be used when working with lye.

# My neighbor has found bee droppings on their car

## CAUSE

For hygienic reasons, bees usually relieve themselves outside the hive. If weather permits, they will fly some distance away from the hive to do this. Trees and other barriers can force bees to develop "restroom" pathways. If the neighbor's automobiles or buildings are beneath the bees' flight corridor, fecal spotting will occur.

## SOLUTION

A few spots on a car bonnet are not much of an issue and are hardly noticeable, but as the spots accumulate, they become apparent. Initially, unsuspecting neighbors feel that small birds or large insects are the cause of the mysterious spots. Invariably, these people will get the word that this is the result of bees taking cleansing flights.

The only solution is to reduce the number of colonies being kept at a particular location. Otherwise, there is no way to force the bees to fly in different directions. Fecal spotting is one of urban beekeeping's big four challenges (along with stinging, water foraging, and swarm settling). These four issues can cause cantankerous interactions with nearby residents.

The spots are durable and will require effort to wash away. Many products are available for washing and cleaning automobiles. Select one that seems appropriate from a simple car-washing view. Mix the solution according to instructions and wet the car, taking care to wet the fecal spots. After soaking for ten minutes or so, wash the car using a microfiber towel. If some spots are persistent, use a bug- and tar-removal product. These spots are easier to remove if they are washed before hardening. Keeping the automobile waxed makes it easier to wash the spots away.

# Spots of beeswax are hard to remove

## CAUSE

**In the routine procedures of beekeeping and honey processing, pieces of beeswax invariably end up in the wrong place. Since wax and propolis are both highly water resistant, fragments of these can become affixed and are remarkably difficult to remove.**

## SOLUTION

The accumulation of these spots and residues is not so much a problem as a typical aspect of beekeeping, candle making, or batiking. If burr combs are scraped from hive equipment in the field, it is not uncommon for the beekeeper to step on these discarded bits. In most instances, much like spent chewing gum on a shoe sole, the wax blob must be scraped away.

When removing beeswax from clothing, let the wax harden. Then, use a dull knife to scrape away as much wax as possible. After scraping, if the wax is pigmented, soak the wax spot in a stain remover. Then put an absorbent paper towel beneath the wax spot and touch the spot with a hot iron. The towel will absorb any residual wax.

During spring and early summer months, sticky propolis is notorious for adhering to hands, bee gloves, hive tools, and clothing. Probably the worst propolis contamination situation is on camera equipment. In most instances propolis on clothes, tools, and cameras can be removed with isopropyl alcohol. For small propolis removal jobs, alcohol-based lens cleaning wipes can be used. Be aware that with both pigmented beeswax and propolis, a stained area may remain after the spot has been cleaned.

For removing propolis from hands, soak your hands with a common alcohol-based hand sanitizer. While hands are still wet, add a large dollop of pumice-based hand cleaner to the mix and scour thoroughly. A nailbrush is useful for clearing propolis from tight spots around cuticles.

# 100 Rendering beeswax over a flame is a fire hazard

## CAUSE

**Beeswax, in any form, is highly flammable. Not only is fire a hazard, but hot wax can cause painful burns.**

## SOLUTION

Simply not using open fire in the melting process is the easiest solution. However, if heat from open fire must be used, set up outside in a safe area away from buildings and human or animal activity. Alternatively, if a gas stove must be used inside, double boilers (one with wax, the other with water) can be used to melt wax over a flame. It should be constantly monitored. If only small amounts of wax need to be melted, put the wax combs or cappings in a pan of water and apply heat. At approximately 150°F (65.5°C), wax will melt and float on the water. After cooling, the floating solidified wax can be removed and used in any way desired. Since wax moth larvae cannot complete their life cycle in rendered wax, this process protects this product.

A wide variety of either homemade or commercially manufactured wax melters are available for the hobby beeswax candle maker. These units safely melt the wax at the minimum melting temperature and hold it at that point. Retired "Crock Pots" are frequently given a second life as a wax-melting device.

Another option is to use solar wax melters. These glass-covered boxes sit in full sunlight. On bright, warm days, this device easily reaches high enough heat to melt beeswax. They are cheap to operate but notoriously inefficient and leave a good deal of unmelted wax. Additionally, wax collected from a solar melter is partially bleached. Depending on its future use, yellow beeswax that has been bleached white by sunlight may or may not be a good thing.

## SLUMGUM

Slumgum, found on the bottom of the hardened wax cake, is composed of cocoons and detritus from the colony nest and is usually dark brown to black. Typically, the quantity of wax contained in slumgum is high, but not always. Cappings wax is cleaner and produces a smaller amount of slumgum than dark wax. Commercial companies can extract wax from this little-known hive product.

✻ *For pouring only a few candles, the candle maker is using a temperature controlled, double-walled heating vessel that provides a controlled heat source. No open flames are involved. Other than occasional minor hand burns, no safety issues are apparent.*

# GLOSSARY

**ABSCONDING**
The complete abandonment of a hive, usually caused by starvation or by pests.

**ALARM PHEROMONE**
A chemical that is produced by worker bees to incite other bees to colony defense.

**AMERICAN FOULBROOD (AFB)**
A bacterial infection of honey bee larvae caused by *Paenibacillus* larvae. Spores cause long-term contamination of equipment.

**APIARIST**
Another term for beekeeper.

**APIARY**
A bee yard location.

**APICULTURE**
The science and craft of keeping bees.

**BANKING QUEENS**
The beekeeper management technique of holding multiple caged queens for later use in other colonies.

**BEE BREAD**
A mixture of pollen and honey prepared by house bees and used to feed developing brood.

**BEE ESCAPES**
Any of several styles of devices that direct bees in a one-way fashion; for example, removing bees from supers or from a dwelling.

**BEE PARASITIC MITE SYNDROME (BPMS)**
A secondary viral infection found in some colonies that have Varroa mite infestations.

**BEE SPACE**
The basic tenet of modern beekeeping hive equipment. Any gap less than ¼in (0.6cm) bees will fill with propolis; ¼–⅜in (0.6–1cm) bees will leave open; greater than ⅜in (1cm) bees will fill with comb.

**BEE SUIT**
Protective clothing, including facial protection, worn by beekeepers when working beehives.

**BEESWAX**
The basic building material of the bee colony, used for brood production and foodstuff storage.

**BROOD**
All stages of developing bees—eggs, larvae, pupae—found in the brood chamber.

**BROOD CHAMBER**
The area of the hive where young bees are produced. Usually near the bottom of the hive.

**BROOD REARING**
The process of producing adult bees, from egg to adult.

**BURR COMB**
Also called brace comb. Built by bees to bridge combs within the hive and on the tops of frames.

**CAPPED BROOD**
The pupal stage of bee development. These cells are sealed with a mixture of wax and propolis.

**CAPPINGS**
Thin, white new beeswax used to cover cells filled with new honey.

**CASTE SYSTEM**
The three types of bees within the hive: queen, workers, and drones.

**CHALKBROOD**
A fungal disease of bee larvae caused by *Ascosphaera apis*. Larvae appear to be white, hard mummies.

## CHUNK HONEY
A type of honey packaging in which a piece of honey is put into a jar along with liquid honey.

## CLEANSING FLIGHT
A flight during which the bee evacuates its rectal contents away from the hive.

## CLUSTER
A compact, heat-producing wintering mass of bees.

## COLONY
The total parts of the nest and its individuals that is housed in a hive.

## COMB
The basic component of the colony. It is made from beeswax.

## COMB HONEY
Honey that is never extracted from the comb, and is produced and sold within basswood boxes or round plastic containers.

## CROSS COMB
Naturally built comb that is nearly impossible to remove from the colony. Such comb is intertwined and not in straight combs. Colonies are said to be "cross combed."

## CRYSTALLIZED HONEY
Honey in which the excess sugar has become solid. Heating drives the honey back into solution.

## DEARTH
A seasonal period of time when there is no nectar, no pollen, or neither, is available for bee foraging.

## DEEP SUPER
The largest storage available to beekeepers for honey storage, commonly 9½in (24cm) deep. They are also routinely used for brood chambers.

## DIVIDING COLONIES
The process of taking part of an established hive, giving it a new queen, and developing another colony from the parent colony.

## DRAWN COMB
A completed comb, as opposed to foundation.

## DRONE
The male bee in the colony. The main function of drones is to mate with new queens.

## DRONE COMB (OR FOUNDATION)
Comb (or foundation) having about four cells per linear inch, whereas worker cells have about five cells per linear inch.

## DRONE LAYER
A defective queen that produces only drones as her offspring.

### DWINDLING
The process of a slow population decline caused by weather conditions or by various hive diseases.

### EGG
Normally produced by queens. The first phase of honey bee development, which lasts for about three days.

### ENTRANCE REDUCER
Various devices used to restrict the hive entrance to a small opening. Commonly used to prevent robber bees entering the hive, or during winter months to reduce heat loss and to keep mice out of the colony.

### EUROPEAN FOULBROOD (EFB)
A bacterial infection of honey bee larvae caused by a bacterium, *Mellitococcus pluton*, and other bacteria as well. Not generally as serious as AFB.

### EXTRACTED HONEY
Honey that is uncapped and spun from the combs, commonly using centrifugal force. This honey is sold as a liquid with no comb.

### EXTRACTOR
A barrel-shaped centrifuge used to extract honey from the combs.

### FEEDER
Any of several styles of devices used to provide sugar syrup or dry sugar to the bee colony.

### FERAL BEES
Wild bees that are not managed by a beekeeper.

### FERMENTATION
The degradation of honey through the action of various yeasts. Honey with moisture higher than 18.6 percent water is prone to fermentation and will turn to a vinegary, sour product.

### FERTILE QUEEN
A queen that has mated with drones and is capable of producing both fertilized and unfertilized eggs.

### FORAGER BEE
A worker bee that goes out to gather water, pollen, nectar, or propolis. Foraging is the last stage of a worker bee's life.

### FOUNDATION
An embossed beeswax sheet or beeswax-coated sheet used by the bees as a template for comb construction.

### FRAME
Usually made of wood, fitted with foundation, and is used to support beeswax comb. May also be made of plastic.

### HIVE
The colony's physical home.

### HIVE STAND
Any of several designed structures used to support the hive off the ground.

### HIVE TOOL
Commonly called a window-opener in the construction trade. Used to pry hive appliances apart and to scrape away propolis and wax when necessary.

**HONEY**

The food produced by the enzymatic inversion of long chain sugars in nectar and the evaporation of water to less than 18.6 percent.

**HOUSE BEE**

The stage of an adult bee's life when it works on building combs and cleaning the hive.

**INSERT**

Inserts are more correctly called foundation inserts. These are rigid plastic sheets that have been embossed with a hexagonal pattern that bees use to begin comb construction. The heavy sheets provide a sturdy comb that can withstand the rigors of honey extraction.

**LANGSTROTH HIVE**

The early hive design used today in simplified form that was developed by L. L. Langstroth, a dedicated beekeeper from Pennsylvania.

**LARVA (PLURAL: LARVAE)**

The grub or worm stage of honey bee metamorphosis. The second stage in the growth sequence: egg, larva, pupa, adult.

**LAYING WORKER**

Workers that have not been hormonally suppressed by pheromone secretions from a fertile queen. Only drones are produced by laying workers.

**MARKED QUEEN**

A queen that has had a distinctive mark, usually enamel paint, put on her thorax by the beekeeper in order to make her easier to find; may also indicate her age.

**MATING FLIGHTS**

Flights taken by a queen only in the early stages of her life, during which she mates. She never mates again.

**METAMORPHOSIS**

System of physiological development by passing through specific stages. Honey bees undergo complete metamorphosis having the stages of: egg, larva, pupa, and adult.

**MIGRATORY BEEKEEPING**

Beekeepers who move their colonies from site to site for either honey production, pollination, or both.

**MOVEABLE FRAME**

A suspended frame that can be removed for inspection or for management reasons.

**NASANOV GLAND**

A scent gland used by bees for navigation and orientation, especially when swarming or foraging.

**NECTAR**

A sugar-solution reward produced by plant nectaries, and offered by some plants to pollinators. Honey bees use nectar to produce honey.

**NECTAR FLOW**

Seasonal period when plants are producing nectar that bees can convert to honey. Sometimes called a honey flow.

## NUCLEUS HIVE
Generally just a small hive with fewer frames than a full-sized hive, or with smaller frames than common frames. Sometimes called a nuc.

## NURSE BEES
Individual bees at a stage in their lives where they are responsible for feeding young bees.

## OBSERVATION HIVE
A hive with either glass or Plexiglas panels for viewing the internal workings of a bee colony. Many styles exist.

## OUTER COVER
The outer top of the bee colony. The inner cover is underneath this cover.

## PHEROMONES
Chemical secretions that direct bees to perform various functions or behaviors. Used by bees extensively.

## POLLEN
The male portion of a blossom required by a specific plant to produce seeds and fruit. Collected by bees and used as a protein source.

## POLLEN BASKET
An improvised receptacle made up of hairs on the rear-most legs of a bee in which pollen is transported back to the colony.

## POLLEN PATTY
Also called a pollen supplement. Made of pollen and other protein-rich foods, this is fed to bees in order to build up colony populations in spring.

## POLLEN TRAP
A device that scrapes the pollen loads from returning forager bees and collects it in a storage container.

## POLLINATION
Fertilization accomplished by appropriate pollen being deposited on the pistil of the blossom. Plants use many diverse pollinating agents to transfer pollen, including bees.

## PROPOLIS
Bee glue or bee caulking. Made of resins collected from trees and buds and used to seal the colony and as a component in brood cappings. Thought by some to have antibiotic characteristics.

## PUPA (PLURAL: PUPAE)
The third stage of complete metamorphosis. Egg, larva, pupa, and adult.

## QUEEN
The chemical and genetic center of a colony. She produces all the brood within the colony and is responsible for specific pheromone production.

## QUEEN CAGE
A cage used to contain a queen for shipping or introduction. Many designs exist.

## QUEEN CELL CUP
A queen cell that is not in use. A cup becomes a cell once an egg is placed in it.

## QUEEN CELL
A peanut-shaped cell used to incubate a new queen. Three types are: emergency, supersedure, and swarm cell.

## QUEEN EXCLUDER
A grid with spacing large enough for worker bees to pass through but restrictive to both queens and drones. Used to keep brood out of the honey supers.

## QUEENRIGHT
A colony that has an established queen is said to be queenright.

## QUEEN SUBSTANCE
A pheromone produced by the queen that inhibits the production of new queens.

## RENDERING WAX
Melting wax cappings and comb residue into wax cakes.

## REQUEENING
The process of replacing a queen within a specific colony.

## REVERSING DEEPS
The management process of exchanging positions of brood chambers and, later in the season, changing the position of supers.

## ROBBING
The behavior of foragers from one colony taking stores from another, usually weaker, colony. Normally happens when no nectar is available in the field.

## ROYAL JELLY
A food rich in hormones and protein, fed by nurse bees to developing queens throughout their development. Worker bees are only fed royal jelly for the first few days of their lives.

## SCOUT BEE
An individual bee searching for a new nest cavity or for food and water resources.

## SETTLING TANK
Usually a stainless steel tank in which honey is allowed to sit for a few days in order for extraneous materials, such as air bubbles, wax, and bee body parts, to float to the surface.

## SHALLOW SUPER
Normally a super with a depth of about 4½in (11cm).

## SLUMGUM
The dark, wax residue remaining after wax has been rendered.

## SMALL HIVE BEETLE (SHB)
The small hive beetle, *Aethina tumida*, is a small, nearly square reddish-brown beetle. The adults and larvae attack honey bee hives and stored honey in combs.

## SMOKER
A beekeeper appliance used to suppress the defensive behavior of guard bees. It is useful in reducing the stinging response of bees.

## SOLAR WAX MELTER
A box with a glass lid that uses solar energy to melt wax. Though cheap to operate, it is not very efficient.

## SPLIT
A small colony made by dividing up a larger colony into several parts. Normally, each split is provided with a new queen and some brood and food reserves.

## STING (OR STINGER)
The defensive weapon of the honey bee.

## SUPER
Additional space supplied by the beekeeper for the bees to store surplus honey.

## SUPERSEDURE
The natural colony process of replacing a failing honey bee queen.

## SUPERING
The management process of adding supers to a hive.

## SWARM
Colony fission. A colony splits, roughly in half, and the split finds a new nest cavity and begins a second colony. The old queen goes with the swarm.

## TOP BAR
The uppermost part of a frame onto which both end bars are attached and from which the comb is suspended.

## TWO-QUEEN SYSTEM
A honey production system using two queens to build up population faster. Normally, one queen is removed just before the nectar flow begins.

## UNCAPPED BROOD
Either the egg or larval stages of bee brood. Also called open brood.

## UNCAPPING KNIFE
Any kind of knife used to cut wax cappings from honeycombs. Usually heated in some way. May be highly mechanized.

## VARROA MITE
A large external parasitic mite (*Varroa destructor*) that attacks both brood and adult bees. If left untreated, most colonies will usually die from mite feeding damage.

## VENOM
Complex chemical produced by glands near the sting of the honey bee that causes pain when injected into potential enemies of the colony.

## VIRGIN QUEEN
A young adult queen that has not taken mating flights.

## WASHBOARDING
A scouring movement, sometimes called rocking, in which bees appear to clean or polish the landing board.

## WAX BLOOM
A crystalline, powdery substance that appears on stored wax. It causes no harm and cannot readily be avoided.

## WAX MOTH (GREATER)
A pest of combs. The larvae of *Galleria mellonella* tunnel through comb while searching for protein and carbohydrates. Combs are destroyed.

## WORKER BEE
The common honey bee that maintains the hive, gathers water and nectar, and pollinates plants. Though diploid, workers are sterile.

# INDEX

# FURTHER RESOURCES

## BOOKS

Crane, Eva. *Honey, A Comprehensive Survey*. London, UK: Heinemann, 1975.

Graham, Joe. *The Hive and the Honey Bee*. Hamilton, IL: Dadant & Sons, 1992.

A. I. Root Company. *The ABC & XYX of Bee Culture*. A. I. Root Company, Medina, OH: A. I. Root Company, 2007.

Laidlaw, Harry H. Jr. and Page, Robert E. Jr. *Queen Rearing and Bee Breeding*. Cheshire, CT: Wicwas Press, 1997.

Morse, Roger A. and Kim Flottum. *Honey Bee Pests, Predators and Diseases*. Medina, OH: The A.I. Root Company, 1997.

Winston, Mark. *The Biology of the Honey Bee*. Cambridge, MA: Harvard University Press, 1987.

Tew, James E. *Backyard Beekeeping*. Tuscaloosa, AL : Alabama Cooperative Extension System, 2004.

## ONLINE RESOURCES

Agricultural Research Service, US Department of Agriculture
www.ars.usda.gov

Apiservices
http://beekeeping.com

BeeBase
www.fera.defra.gov.uk

Extension Beekeeping Information
www.extension.org/bee_health

National Honey Board (USA)
www.honey.com

James Tew
www.onetewbee.com

# IMAGE CREDITS